1 MONTH OF
FREE
READING

at

www.ForgottenBooks.com

By purchasing this book you are eligible for one month membership to ForgottenBooks.com, giving you unlimited access to our entire collection of over 1,000,000 titles via our web site and mobile apps.

To claim your free month visit:
www.forgottenbooks.com/free898384

ISBN 978-0-266-84655-0
PIBN 10898384

GEOLOGY OF THE EDDYVILLE, STONEFORT, AND CREAL SPRINGS QUADRANGLES, SOUTHERN ILLINOIS

GEOLOGY OF THE EDDYVILLE, STONEFORT, AND CREAL SPRINGS QUADRANGLES, SOUTHERN ILLINOIS

W. John Nelson, Joseph A. Devera, Russell J. Jacobson,
Donald K. Lumm, Russel A. Peppers, Brian Trask,
C. Pius Weibel, Leon R. Follmer, and Matthew H. Riggs
  Illinois State Geological Survey

Steven P. Esling, Elizabeth D. Henderson, and Mary S. Lannon
  Southern Illinois University, Carbondale, IL

BULLETIN 96   1991

ILLINOIS STATE GEOLOGICAL SURVEY
Morris W. Leighton, Chief
615 East Peabody Drive
Champaign, IL 61820

CONTENTS

Figures

Figures

Tables

Plate

ABSTRACT

The Eddyville, Stonefort, and Creal Springs 7.5-Minute Quadrangles are in southern Saline and Williamson and northern Johnson and Pope Counties, Illinois. Most of the study area is within the Shawnee Hills section of the Interior Low Plateaus physiographic province; the northern edge is in the Central Lowlands province. Sedimentary rocks of Mississippian and Pennsylvanian age crop out in the Shawnee Hills, and Pleistocene glacial deposits cover Pennsylvanian bedrock in the Central Lowlands.

Exposed Mississippian rocks are assigned to the Clore and Degonia Formations and the Kinkaid Limestone, all of upper Chesterian age. The Clore contains 95 to 155 feet of shale, sandstone, and limestone, of which the upper 30 feet is exposed. The Degonia consists of 20 to 64 feet of shale, siltstone, and very fine-grained sandstone. The Kinkaid, 85 to 230 feet thick, consists of limestone, shale, claystone, and sandstone. A prominent unconformity separates the Kinkaid from Pennsylvanian strata.

The oldest Pennsylvanian formation is the Caseyville Formation, of Morrowan age. About 200 to 450 feet thick, the Caseyville consists primarily of quartz-arenitic sandstone. Four members are recognized: the Wayside (oldest), Battery Rock Sandstone, Drury, and Pounds Sandstone. The Battery Rock and the Pounds are thick (30 to 120 ft), cliff-forming, widely traceable, thick-bedded sandstones that commonly contain quartz granules and pebbles. The poorly exposed Wayside Member is composed of gray to black, silty shale and siltstone, very fine-grained, thin-bedded sandstone, and lenses of thick-bedded sandstone. The Drury Member is similar but contains thin, discontinuous coal beds. The Caseyville represents a variety of depositional environments, ranging from estuary or embayment (shale with goniatites) to swamp (coal and rooted claystone), in an overall deltaic setting.

The overlying Abbott Formation, of late Morrowan and Atokan age, thickens eastward from 220 to 400 feet across the study area. The Abbott consists of numerous sandstone bodies, more lenticular than those of the Caseyville, interbedded, and intertonguing with gray to black shale, siltstone, and local coal. Abbott sandstones are mineralogically less mature than Caseyville sandstones and generally lack quartz pebbles. The proportion of sandstone in the Abbott increases eastward. Shale and siltstone in the Abbott locally contain abundant trace fossils, some indicating marine sedimentation; some marine body fossils have also been found. The Abbott is interpreted as consisting mostly of subaqueous deposits of a series of deltas that prograded westward and southwestward.

The lower Desmoinesian Spoon Formation thickens eastward from 200 to 300 feet. In comparison with the Abbott, the Spoon contains less sandstone, more shale, and thicker and more widespread coal and limestone beds. The Spoon represents deltaic and shallow-marine sedimentation in an increasingly stable regime.

The Carbondale Formation, exposed in a small part of the Eddyville Quadrangle, is the youngest bedrock in the study area. The Carbondale contains regionally continuous, thick coal and black fissile shale, and displays pronounced cyclicity.

The Glasford Formation, an Illinoian glacial till, covers the northwestern Creal Springs Quadrangle. Adjacent valleys contain Illinoian ice-margin deposits, which we have correlated with the Teneriffe Silt. Sediments of Wisconsinan slackwater lakes in the northern part of the study area are termed Equality Formation. Uplands throughout the area are mantled with residuum (Oak formation), colluvium (Peyton Formation), and loesses: the Loveland (Illinoian), Roxana (Altonian), and Peoria (Woodfordian). The Holocene Cahokia Alluvium occupies stream valleys throughout the region.

Structurally, the region lies on the southern margin of the Illinois Basin; bedrock strata dip regionally northward. Folds and faults in the report area represent three major episodes of deformation. The first was normal faulting related to rifting that occurred near the end of Precambrian time along the Shawneetown and Lusk Creek Fault Zones at the southeastern corner of the map area. The second was Permian (?) compression, which induced high-angle reverse faulting in the Lusk Creek Fault Zone and detached thrusting and folding in the McCormick and New Burnside Anticlines. The third event was Triassic or Jurassic (?) extension, during which earlier formed reverse faults underwent normal movement and many new normal faults were formed. The McCormick Anticline may have begun to rise during Morrowan and Atokan sedimentation; evidence for the uplift includes paleoslumps, local unconformities, and facies patterns that appear to follow the anticline.

Coal has been extracted in the study area from surface mines and from small drift and slope mines. Coals of the lower Abbott and Spoon Formations, particularly the Delwood and Mt. Rorah Coals, offer prospects to small operators; however, careful exploratory drilling must be done because of the variable thickness and quality of the coals and their limited continuity. Twenty-nine petroleum test holes have been drilled in the study area; all were dry and were abandoned. Chesterian targets on major anticlines have been tested; deeper possibilities are speculative. Fluorspar has been mined from veins along the Lusk Creek Fault Zone, and reserves probably exist at depth there. Limestone, sand, and gravel can be quarried, but poor access and distance to market limit the potential for development.

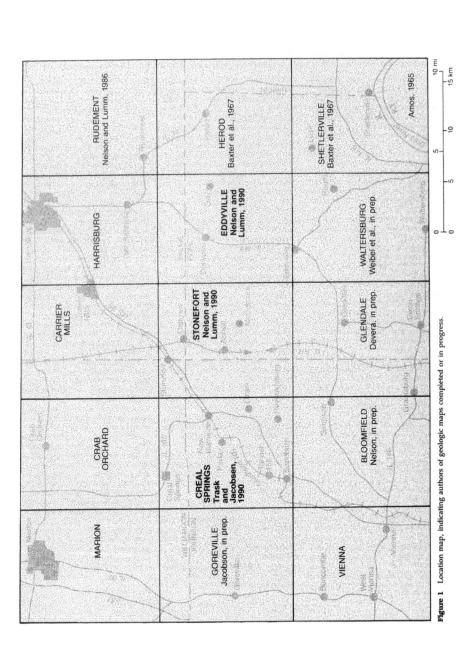

Figure 1 Location map, indicating authors of geologic maps completed or in progress.

INTRODUCTION

This study was conducted as part of the ongoing Cooperative Geologic Mapping Program (COGEOMAP) of the U.S. Geological Survey (USGS) and the Illinois State Geological Survey (ISGS). The USGS initiated COGEOMAP in 1985 as a national program to revitalize geologic mapping. The first cooperative agreement between the ISGS and the USGS, signed in 1985, provided that the ISGS supply personnel and research facilities and the USGS supply technical and financial assistance. Survey staff had mapped part of the Creal Springs Quadrangle before 1985. The initial COGEOMAP agreement called for publication of three quadrangles already mapped (Nelson and Lumm 1986a,b,c) and for mapping an additional twelve 7.5-minute quadrangles.*

The Eddyville, Stonefort, and Creal Springs Quadrangles were selected as the initial COGEOMAP quadrangles for several reasons: (1) the proximity of these quadrangles to quadrangles already mapped or being mapped; (2) the desire to follow up challenging structural and stratigraphic problems that had been identified during previous mapping (for instance, several fault zones and anticlines that have received little attention occur in the quadrangles); (3) the exposures of the problematic lower Pennsylvanian succession (below the Carbondale Formation) in the quadrangles, and the existence of type localities for many named units in or near the study area; and (4) the known existence but poor delineation of coal resources in the quadrangles, and the unsuccessful oil exploration conducted in the region.

Location and Climate

The Eddyville, Stonefort, and Creal Springs Quadrangles are in southwestern Saline, southeastern Williamson, northeastern Johnson, and

northwestern Pope Counties in southeastern Illinois (fig. 1). The study area is rural; the largest towns are Creal Springs (population 845), Stonefort (population 316), New Burnside (population 276), and Eddyville (population about 100). Most of the land is privately owned, but the U.S. Forest Service holds surface and mineral rights in large tracts in the Shawnee National Forest, especially in the Eddyville and Stonefort Quadrangles. Topographic maps showing Forest Service lands and trails are available from the Forest Service office in Harrisburg, Illinois.

Southern Illinois has a warm temperate climate. Rainfall averages 42 inches yearly; spring is the wettest season and fall the driest (Miles and Weiss 1978). Most of the geologic mapping was conducted in the cool months (from late October to early May) when vegetation presented less of a problem.

Topography

Most of the study area is in the Shawnee Hills section of the Interior Low Plateaus physiographic province (Leighton et al. 1948, Horberg 1950). The Shawnee Hills are a strip of rugged bedrock hills between the drift-covered Central Lowlands to the north and the Mississippi Embayment to the south. A southern belt of the Shawnee Hills—most of which lies south of the study area—is underlain by Mississippian limestone, shale, and sandstone that produce a gently to moderately rolling topography and broad, alluviated stream valleys. The northern belt of the Shawnee Hills, underlain by sandstone and shale of Pennsylvanian age, has a more rugged topography. The Pounds Escarpment, composed of massive, conglomeratic, basal Pennsylvanian sandstones, is near the southern edge of the map area. The area north of the Pounds

Escarpment is a moderately rolling upland cut by many deep ravines.

Drainage in the study area is relatively youthful: the region has been subjected to erosion for a relatively short time. Small streams are well adjusted to bedrock structure, but larger ones cut across structures and follow incised meanders. Most larger streams have steep sides and flat, alluviated bottoms. Small ravines generally have a V-shaped profile.

Part of the northwestern Creal Springs Quadrangle is mantled with glacial drift, and in this region the topography is gently rolling. Broad, level, swampy areas along streams in the northernmost parts of all three quadrangles are underlain by glaciolacustrine sediments. These areas are part of the Mt. Vernon Hill Country, a division of the Central Lowlands (Leighton et al. 1948).

Bedrock outcrops in the study area are found mainly along ravines. Sandstone forms bluffs or ledges along valley walls and crops out extensively along the beds of small to medium-sized streams. Some Pennsylvanian sandstones are resistant enough to produce cliffs away from streams. Prominent hogbacks of these sandstones occur along anticlines and faults. Exposures of shale and siltstone are small and generally confined to cutbanks of active streams. Shale, siltstone, and limestone are grossly underrepresented in natural exposures because of the humid climate and extensive soil and vegetation. Railroad cuts, roadcuts, and exposures in mines provide more complete views of the strata but are few and far between.

Geologic Setting

The study area is on the southern margin of the Illinois Basin, a structural and depositional basin (fig. 2). The Illinois Basin contains sedimentary rocks ranging in age from Cambrian to earliest Permian. The present

*In 1986, geologic maps of the Shawneetown, Equality, and Rudement Quadrangles were published in the new ISGS series, Illinois Geologic Quadrangles (IGQ). The accompanying report was published as Circular 538 (1987). The Eddyville, Stonefort, and Creal Springs IGQs discussed in this report were published in 1990 as IGQ-4, IGQ-5, and IGQ-6.

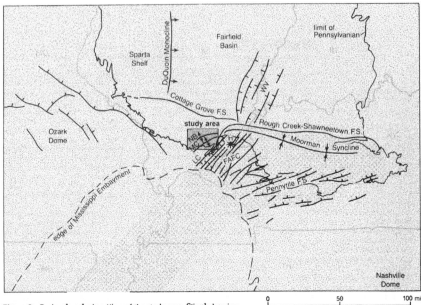

Figure 2 Regional geologic setting of the study area. Stippled region represents Late Precambrian-Early Cambrian rift system (Reelfoot Rift and Rough Creek Graben). FAFC=Fluorspar Area Fault Complex; H = Hicks Dome; LC = Lusk Creek Fault Zone; MC = McCormick Anticline; NB = New Burnside Anticline; WV = Wabash Valley Fault System. (Modified from King and Beikman 1974).

configuration of the basin is largely the result of uplift on the south (Pascola Arch) that developed middle Pennsylvanian time. The thickest preserved Paleozoic succession in Illinois (about 15,000 ft) is at the center of the Fairfield Basin. Thicker sedimentary rocks—possibly 25,000 feet—are found in western Kentucky, south of the Rough Creek-Shawneetown Fault System.

The southern part of the Illinois Basin is extensively faulted (fig. 2). A swarm of northeast-striking, high-angle faults crosses southeastern Illinois and the adjacent part of Kentucky. The name Fluorspar Area Fault Complex has been applied to these faults because many of them contain commercial vein deposits of fluorite (Grogan and Bradbury 1968). The Lusk Creek Fault Zone, which crosses the southeastern corner of the Eddyville Quadrangle, marks the northwest edge of the Fluorspar Area Fault Complex. The Rough Creek-Shawneetown Fault System branches off the Lusk Creek Fault Zone at the

eastern edge of the Eddyville Quadrangle and strikes north-northeast. About 12 miles northward the fault system abruptly curves to the east and continues more than 100 miles into Kentucky. The Rough Creek-Shawneetown System contains high-angle reverse and normal faults and bears evidence of several episodes of movement (Nelson and Lumm 1984). The Cottage Grove Fault System, a right-lateral strike-slip system (Nelson and Krausse 1981), crosses southern Illinois east to west just north of the study area.

Previous Studies

The first known report covering part of the study area was a description by Cox (1875) of the geology of Saline and Gallatin Counties. Brokaw (1916) discussed geology and oil possibilities of southern Illinois and mapped structure, including the McCormick and New Burnside Anticlines. Cady (1926) mapped the areal geology of Saline County. Weller (1940), who described and classified Pennsylva-

nian rocks and speculated on the origin of structures in and near the present study area, also discussed oil possibilities in southern Illinois. Smith (1957) mapped strippable coal reserves and refined the stratigraphy of Pennsylvanian rocks in and around the study area. Potter (1957) examined deformed Pennsylvanian rocks in railroad cuts in the Stonefort Quadrangle and theorized on the cause of the deformation. His report contains a geologic map of a small part of the Stonefort Quadrangle.

No geologic maps covering more than a fraction of the study area have been published at a scale larger than 1:500,000. Klasner (1982, 1983) mapped the geology of part of the current study area at a scale of 1:24,000. His purpose was to assess the mineral potential for two proposed "wilderness" areas in the National Forest. The 1982 map covers part of the southeastern Eddyville Quadrangle; the 1983 map includes part of the west-central Eddyville and east-central Stonefort Quadrangles.

Table 1 COGEOMAP cored test holes in the study area

Hole	Location		TD (ft)	Strata penetrated
			Creal Springs Quadrangle	
C-1	NE NE NW	35-11S-3E	110	Middle and lower Abbott Formation (fig. 10)
C-2	NE NW SW	23-11S-3E	83	Clifty Creek sandstone, olive shale (fig. 16)
C-3	SW NE NE	25-11S-3E	151	Middle Abbott Formation including marine lime-stone; Tunnel Hill Coal Bed (fig. 10)
C-4	SE NE SE	29-10S-3E	101	Sub-Davis sandstone, Carrier Mills Shale Member, Wise Ridge Coal Bed, Mt. Rorah Coal Bed, Creal Springs Limestone Member (fig. 21)
C-5	NE SE SE	36-11S-3E	256	Middle and lower Abbott Formation; Tunnel Hill and Reynoldsburg Coals (fig. 10)
C-6	NW NW SE	15-11S-4E	266	Middle and lower Abbott Formation including marine limestone and Tunnel Hill Coal Bed (fig. 10)
C-7	SW SW NE	5-11S-4E	156	Golden sandstone, Delwood Coal Bed, Clifty Creek sandstone (fig. 20)
C-8	NW NE NE	26-11S-3E	81	Clifty Creek sandstone, olive shale (fig. 16)
			Eddyville Quadrangle	
E-1	SE NE SE	33-10S-6E	201	Golden sandstone, Mitchellsville Limestone Member, Delwood Coal Bed, Murray Bluff Sandstone Member (fig. 20)
E-2	NE NE NW	8-11S-6E	141	Middle and lower Abbott Formation (fig. 10)
E-3	SE NE NE	9-11S-6E	141	Middle Abbott Foramtion (fig. 16)
E-4	SW SW NE	32-11S-6E	181	Pounds Sandstone Member, Drury Member, Battery Rock Sandstone Member (fig. 8)
			Harrisburg Quadrangle	
H-1	SW NW SE	20-10S-4E	300	Sub-Davis sandstone, Carrier Mills Shale Member, Stonefort Limestone Member, Wise Ridge and Mt. Rorah Coal Beds, Mitchellsville Limestone Bed, Delwood Coal Bed, Murray Bluff Sandstone Member (fig. 21)
			Stonefort Quadrangle	
S-1	SW NE SE	25-10S-4E	196	Sub-Davis sandstone, Carrier Mills Shale, Stone-fort Limestone, Wise Ridge and Mt. Rorah Coal Beds, Creal Springs Limestone Member (fig. 21)
S-2	SW NE SE	29-10S-5E	265	Delwood Coal Bed, unnamed coals; "golden sand-stone" (fig. 20)
S-3	SE SE NE	33-10S-5E	−136	Golden sandstone, Oldtown Coal Bed, Murray Bluff Sandstone Member (fig. 20)
S-4	NW SE NW	25-11S-4E	346	Murray Bluff Sandstone Member, olive shale, lower Abbott sandstone, Tunnel Hill and Reynoldsburg Coal Beds, uppermost Pounds Sandstone Member (fig. 16)
			Waltersburg Quadrangle	
W-3	NW NE NE	9-12S-5E	280	Basal Caseyville Formation, Kinkaid Limestone, Degonia Formation and uppermost Clore Formation

Accompanying the maps are discussions of the geology, geochemistry, and mineral potential for the two areas. Klasner's maps show less detail than ours, and some of his interpretations, particularly of faulting, are not supported by our findings.

Unpublished geologic maps, manuscripts, and field notes of parts of the study area are available for inspection in the ISGS library map room.

The region directly east of our study area has been mapped geologically at a scale of 1:24,000 (fig. 1). The quadrangles constituting the Illinois part of the fluorspar district were mapped by Baxter et al. (1963, 1967) and Baxter and Desborough

(1965). Bedrock geology immediately north of the fluorspar district was mapped by Nelson and Lumm (1986 a,b,c). A preliminary geologic map of the quadrangles directly south of the study area was published by Weller and Krey (1939). Lamar (1925) mapped the Carbondale Quadrangle (scale, 1:62,500), which is west of the

Creal Springs Quadrangle. Quadrangles immediately south and west of the present study area are currently being mapped at a scale of 1:24,000 as part of COGEOMAP (fig. 1).

Numerous studies have been published on selected aspects of the geology in and near the Eddyville, Stonefort, and Creal Springs Quadrangles. The most important stratigraphic studies are by Weller (1940), Siever (1951), Kosanke et al. (1960), and Peppers and Popp (1979). The sedimentology and petrology of Pennsylvanian rocks were examined by Potter (1957, 1963), Potter and Glass (1958), and Siever and Potter (1956). Structural geology and igneous rocks of the region were covered by Weller et al. (1952), Stonehouse and Wilson (1955), Clegg and Bradbury (1956), and Nelson and Lumm (1984).

Method of Study

This report is based principally on the geologic study of bedrock outcrops. Jacobson and Trask, assisted by Lumm, Nelson, and Weibel,

mapped the Creal Springs Quadrangle from the fall of 1979 through the spring of 1987. The Eddyville Quadrangle was mapped by Nelson and Lumm, assisted by Devera, Weibel, and Stephen K. Danner, from the fall of 1984 through the spring of 1986. Lumm and Nelson mapped the Stonefort Quadrangle in 1986.

R. A. Peppers conducted palynological studies of coal samples collected during the mapping, and invertebrate fossils were identified by Devera, Jacobson, and Rodney D. Norby.

All available well records for the study area were examined. Logs of boreholes are filed for public inspection in the ISGS Geologic Records. Records include drillers' logs, sample studies, geophysical logs, and very limited core data.

During the mapping program 17 stratigraphic test holes were drilled in or directly adjacent to the study area (table 1; plate 1). All of the test holes were cored continuously, and geophysical logs were obtained for most of the holes. Total depths of

COGEOMAP holes range from 81 to 291 feet. Records of these borings are filed in the ISGS Geologic Records.

Field notes from previous studies by ISGS geologists were gleaned for useful information. These notes include descriptions of outcrops that are now covered by colluvium or otherwise lost to study.

Aerial photography in the study area was of limited value for the geologist. Most aerial photography in southern Illinois is done during the growing season for use in agricultural studies. Few details of bedrock geology can be discerned through the dense vegetation that covers most of the study area during the summer.

Field notes from this study are number-keyed to locations marked on topographic maps and kept on public file in the Map Room of the ISGS library. In this publication, reference is occasionally made by number to specific field notes that can provide the reader with additional detail.

6

STRATIGRAPHY OF THE BEDROCK

W. J. Nelson, J. A. Devera, R. J. Jacobson, D. K. Lumm,
R. A. Peppers, C. B. Trask, and C. P. Weibel

CAMBRIAN THROUGH DEVONIAN SYSTEMS

Strata of the Cambrian, Ordovician, Silurian, and Devonian Systems have been penetrated in several deep wells in and near the study area (table 2). The general succession in these wells is similar to the succession in other wells in the southern part of the Illinois Basin (Atherton 1971, Willman et al. 1975, Schwalb 1982). However, two proprietary seismic-reflection profiles viewed by the authors reveal some stratigraphic conditions not previously recognized.

The first profile (fig. 3) through the Eddyville Quadrangle indicates that the Lusk Creek Fault Zone was active during and possibly before deposition of the Upper Cambrian Mt. Simon/Lamotte Sandstone. The Lusk Creek Fault Zone crosses the southeastern corner of the Eddyville Quadrangle. On the seismic profile, as at the surface, the fault zone dips steeply to the southeast. Offset on strata above the Mt. Simon/Lamotte is slight, but the top of crystalline basement appears to be downthrown several thousand feet on the southeastern side of the fault zone. An interval of layered reflectors representing the Mt. Simon/Lamotte, and possibly older strata, correspondingly thickens southeast of the fault zone. The Lusk Creek Fault Zone

lines up with the northwest margin of a geophysically defined major graben within the Reelfoot Rift of southeastern Missouri and northeastern Arkansas (Ervin and McGinnis 1975, Hildenbrand et al. 1977, Howe and Thompson 1984).

The other proprietary seismic reflection profile runs southward from Franklin County, across the Cottage Grove Fault System, to the northern edge of the Mississippi Embayment in Massac County, passing near the western edge of the present study area. This profile showed substantial southward thickening of the Cambrian succession but no indication of growth faulting. The seismic evidence negates Schwalb's (1982) assertion that the Cottage Grove Fault System was active during Cambrian sedimentation.

MISSISSIPPIAN SYSTEM

Mississippian rocks crop out extensively south of the study area and are known from numerous wells (tables 1, 2). The youngest Mississippian strata, representing the Clore, Degonia, and Kinkaid Formations, are exposed in the southeastern Eddyville and southwestern Stonefort Quadrangles. A composite columnar section of Chesterian strata in the study area was prepared from

borehole data (fig. 4). Mississippian formations exposed in the quadrangles are described in the following pages.

Clore Formation

Name and correlation The Clore Formation was named by Weller (1913) for Clore School in Randolph County, southwestern Illinois. Weller (1920) redefined the Clore, excluding the newly named Degonia Sandstone and Kinkaid Limestone. The Clore is readily correlated in outcrop and subsurface throughout southern Illinois on the basis of its lithology, although the proportion of limestone gradually decreases eastward from the type area (Swann 1963, Abegg 1986).

Distribution and thickness The only outcrops of the Clore found in the study area were in a southeast-flowing stream and in rising gullies in NW Section 2, T12S, R4E, near the southwest corner of the Stonefort Quadrangle. The Clore was observed just south of the Eddyville Quadrangle on the west side of Lusk Creek and has been projected a short distance into the Eddyville Quadrangle.

Well records indicate that the thickness of the Clore ranges from 95 to 120 feet in the Eddyville and Stonefort Quadrangles and from 110 to 155 feet in the Creal Springs Quadrangle.

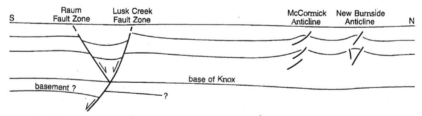

Figure 3 Sketch based on north-south seismic reflection profile through the Eddyville Quadrangle. The profile indicates that (1) Lusk Creek Fault Zone is a listric normal fault that penetrates crystalline basement; (2) Lusk Creek Fault was active during pre-Knox sedimentation; (3) Raum Fault Zone, which forms the southeast margin of the Dixon Springs Graben, is a post-Pennsylvanian structure that intersects the Lusk Creek Fault Zone at depth; (4) the prominent reflector representing the base of the Knox is not deformed beneath the McCormick and New Burnside Anticlines, and therefore these anticlines are detached within upper Paleozoic strata; (5) imbricate faults, most of which dip southward and flatten at depth, underlie the two anticlines.

Table 2 Thickness of strata in selected deep wells in and near the study area

	Streich[a]	Wells et al.[b]	Mohler[c]	Texota[d]	Walters[e]	Farley #1[f]	Boner[g]
Pennsylvanian	418	695	540	0	558	0	525
Mississippian	3,050	3,197	3,350	2,327	3,399	2,140	3,292
Chesterian	1,292	1,052	1,528	867	875?		1,429
Ste. Genevieve	220		340	194	205?		396
St. Louis	430?		222		420?		214
Salem	420?	2,113			292	2,133	352
				1,256			
Ullin	480?		1,175		267		
					843		870
Fort Payne	152?				489		
Springville	48	20	80	0	11		42
Chouteau	8	12	5	10	7	?	7
Devonian	1,762	2,060	350+	745+	1,793	1,531	348+
New Albany	320	345	172	369	385	210	183
Lingle	112		55	88	85		66
						180	
Grand Tower	120	1,715	109	261	209		50
Lower Devonian	1,210		14+	27+	1,114	1,141	49+
Silurian	354	248+			356	336	
Ordovician and Knox	8,222				1,582+	1,581+	
Maquoketa	232				238	237	
Galena					114	124	
	760						
Platteville					576	550	
Joachim and Dutchtown	690				540	595	
St. Peter	190				0	75	
Knox (Camb-Ord)	6,350				124+	6,312	
Cambrian below Knox	1,092+					2,384	
Eau Claire	920					1,160	
Mt. Simon	172+					1,224	
Total Depth	14,942	6,200	4,250	3,172	7,688	14,285	4,165

[a] Texas Pacific, Mary Streich #, SW NW SE, 1-11S-6E, Eddyville Quadrangle.
[b] Texas Pacific, Wells et al. #1, N½ NW NW, 34-10S-6E, Eddyville Quadrangle.
[c] Jenkins, Mohler #1, SW SW NW, 24-11S-3E, Creal Springs Quadrangle.
[d] Texota, King #1, NE SE, 32-10S-7E, Saline County.
[e] Texaco, J. M. Walters #1, SW NE SW, 29-9S-9E, Gallatin County.
[f] Texas Pacific, Farley et al. #1, SE NW NW, 34-13S-3E, Johnson County.
[g] Tunnel Hill Oil Co., Boner #1, W½ SW NE, 30-11S-3E, Johnson County.

Lithology Well logs in the study area show that the Clore has a consistent lithologic sequence correlative with that found in outcrop in adjacent quadrangles (Baxter et al. 1967, Devera, in preparation, Nelson, in preparation, Weibel et al., in preparation). The Cora, Tygett, and Ford Station Members of Swann (1963) can be identified in most logs (fig. 4).

The Cora Member at the base of the Clore is 35 to 70 feet thick and consists of shale and thin limestone beds. The shale is mostly dark gray to greenish gray, silt free, and partly calcareous; the limestone, mostly very fine to fine-grained, argillaceous, and very fossiliferous, is dark gray or brownish gray. Limestone accounts for about one-fourth to one-third of the total thickness of the Cora; most of the limestone is in the upper part of the member in beds a few inches to about 5 feet thick.

The Tygett Sandstone Member ranges from less than an inch to 20 feet thick in the Eddyville Quadrangle and from 15 to 50 feet thick in the Stonefort and Creal Springs Quadrangles. To the east the Tygett consists mostly of gray silty shale and siltstone. Westward, the Tygett becomes sandstone, very fine to fine grained and generally light gray-brown to greenish gray. The sample study of well 15 (table 3) indicates

8

about 8 feet of limestone, lithologically similar to limestone of the Cora Member, near the middle of the Tygett. Thin limestone beds also occur in the Tygett in the adjacent Glendale and Bloomfield Quadrangles (Devera, in preparation, Nelson, in preparation).

The Ford Station Member of the upper Clore consists of 23 to 50 feet of interbedded shale and limestone. Shale is generally predominant, but in a few well logs the Ford Station consists mostly of limestone. Lithologies are similar to those of the Cora Member. The medium to dark gray, fine- to coarse-grained, argillaceous limestone exposed in the southwestern Stonefort Quadrangle is believed to belong to the Ford Station Member. Fossils include the bryozoans *Archimedes*, *Polypora*, and *Rhombopora*; the brachiopods *Composita subquadrata* and *Spirifer sp.*, silicified rugose corals of the genus *Trilophyllites*, and abundant pelmatozoan fragments. The limestone occurs in beds less than 2 feet thick, separated by covered intervals that probably are mostly shale.

Both upper and lower contacts of the Clore Formation appear to be conformable within the study area. The top of the Clore is placed at the top of the highest limestone bed beneath sandstone, siltstone, and silty shale of the Degonia.

Depositional environment The variety of lithologies in the Clore Formation reflects varying deposi-

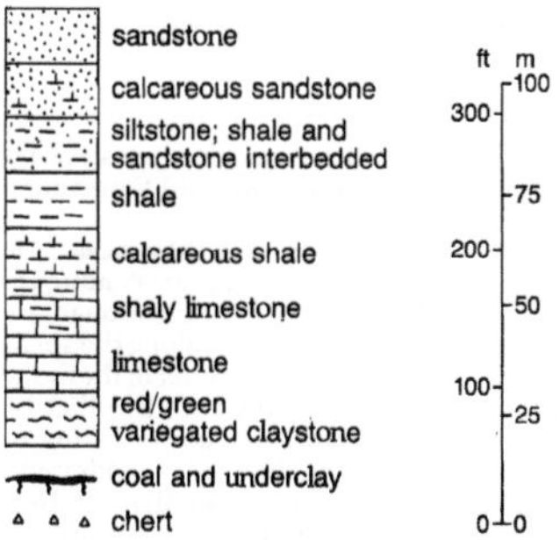

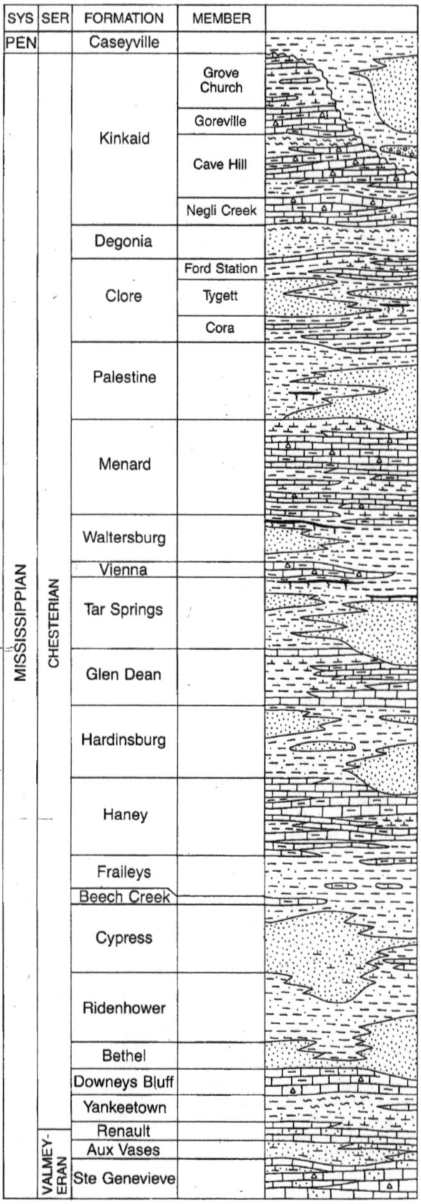

sandstone
calcareous sandstone
siltstone; shale and sandstone interbedded
shale
calcareous shale
shaly limestone
limestone
red/green variegated claystone
coal and underclay
chert

ft   m
    ┌─100
300─┤
    ├─75
200─┤
    ├─50
100─┤
    ├─25
  0─┴─0

Figure 4 Graphic column, based on well data, of the Chesterian Series in the study area (modified from Swann 1963).

SYS	SER	FORMATION	MEMBER
PEN		Caseyville	
MISSISSIPPIAN	CHESTERIAN	Kinkaid	Grove Church
			Goreville
			Cave Hill
			Negli Creek
		Degonia	
		Clore	Ford Station
			Tygett
			Cora
		Palestine	
		Menard	
		Waltersburg	
		Vienna	
		Tar Springs	
		Glen Dean	
		Hardinsburg	
		Haney	
		Fraileys	
		Beech Creek	
		Cypress	
		Ridenhower	
		Bethel	
		Downeys Bluff	
VALMEY-ERAN		Yankeetown	
		Renault	
		Aux Vases	
		Ste Genevieve	

9

Table 3 Oil test holes in the study area

	Operator	Farm, well no.	Location	Year drilled	Total depth (ft)	Deepest unit penetrated
			Eddyville Quadrangle			
1	Carter Oil	Gowdy, #1	NW NW SE 36-10S-5E	1955	3,034	Ullin
2	Texas Pacific	Wells et al., #1	N½ NW NW 34-10S-6E	1977	6,200	Silurian
3	Whitlock et al.	Anthis, #1	SE NW NE 12-11S-5E	1938	1,760	Chesterian
4	Mid-Egypt	Lightfoot, #1	SE SW NE 12-11S-5E	1923?	1,670	Chesterian
5	Mid-Egypt	Stalions, #1	SW NW NE 12-11S-5E	1923	1,760	Chesterian
6	Roy Pledger	Gibson, #1	NW NE NW 13-11S-5E	1956	2,631	St. Louis
7	Texas Pacific	Streich, #1	SW NW SE 2-11S-6E	1976	14,942	Mt. Simon
8	Milo Ditterline	Hart, #1	SE SE NW 14-11S-6E	1957	2,297	St. Louis
9	K-Winn	Headrich, #1	SW NW NE 25-10S-5E	1984	1,530	Hardinsburg
			Stonefort Quadrangle			
10	Ohio Oil Co.	Hancock, #1	SW SE SE 33-10S-4E	1917	3,058	Ullin?
11	Ohio Oil Co.	Bynum, #1	NW NW NE 35-10S-5E	1916	1,281	Chesterian
12	Pierce	Camden, #1	NW NW NW 9-11S-5E	1916	3,075	Salem?
13	Ohio Oil Co.	Gen. Am. Life, #1	NE SE SE 10-11S-5E	1940	1,692	Ste. Genevieve
14	Mid-Egypt	Parsons, #1	NW NE NE 15-11S-5E	1923?	660	Chesterian
15	Gardenheir & Smith	Peoples, #1	SE SE 19-11S-5E	1940	1,160	Menard
16	Claro	Moyer, #1	NE SE NW 29-11S-5E	1941	1,236	Glen Dean
17	Garnier et al.	Peeples #1	NE NE NE 30-11S-5E	1940	2,204	Ste. Genevieve?
18	Higgins	Trammell, #1	SE SE NE 36-10S-4E	1955	2,304	Ste. Genevieve
			Creal Springs Quadrangle			
19	Mitchell & Stanonis	Simmons, #1	SE NE SE 23-11S-3E	1965	1,676	Cypress
20	Mitchell & Hamilton	Giver, #1	NE SE NE 24-11S-3E	1980	2,400	St. Louis
21	Jenkins*	Mohler, #1	SW SW NW 24-11S-3E	1941	4,250	Clear Creek (Devonian)
22	Hardin & Harlow	McCuan, #1	NE NW NE 26-11S-3E	1941	708	Pennsylvanian
23	Witnel-Cunningham		SE SW SE 3-11S-4E		1,508	Chesterian
24	Fletcher Farrar	Horn, #1	SE NW NW 5-11S-4E	1956	2,450	St. Louis?
25	Shure Oil Co.	Evans, #1	SE SE NW 10-11S-4E	1955	2,230	St. Louis
26	Reeves	Gibson, #1	NE NE SW 15-11S-4E	1908	1,560	Chesterian
27	Mitchell & Hamilton	Parson, #1	NE NW NE 16-11S-4E	1980	2,130	St. Louis
28	Wrightsman	Jackson-Whitnal #1	SE SW NW 17-11S-4E	1917	2,002	Ste. Genevieve
29	Monjeb Minerals	Trammel, #1	NE NW SE 27-10S-4E	1983	2,461	Ste. Genevieve

*Also known as Benedum-Trees Oil Co. No. 1 Cavitt. This well was drilled by one operator and later deepened by a second. Its history is unclear.

tional conditions. Because only limited data on depositional environment are available for the Clore in the study area, our discussion is brief; however, Abegg (1986) provides a good overview of the Clore in southern Illinois.

Fossils found in Ford Station Limestone Member in the Stonefort Quadrangle represent forms adapted to low turbulence and soft, muddy substrates. Particularly diagnostic of this setting are specimens of *Archimedes* with fronds attached, spiriferid brachiopods with broad hinge lines, and rugose corals. According to Abegg (1986), the environment was probably a shallow marine shelf

having normal salinity, low rates of sedimentation, and low wave turbulence.

Degonia Formation

Name and correlation Weller (1920) defined the Degonia Sandstone as part of an interval of rock previously assigned to the Clore Formation. In its type area, about 45 miles west of the study area, the Degonia is largely massive sandstone. Eastward towards the study area, the Degonia grades to shale, siltstone, and very fine-grained, thinly bedded sandstone. We define the Degonia Formation in this publication as a siliciclastic interval be-

tween the distinctive limestones of the Clore and Kinkaid Formations.

Distribution and thickness Outcrops of the Degonia are confined to two small areas within the quadrangles studied. One is along the stream at the southwest corner of the Stonefort Quadrangle, and the other is along Lusk Creek at the southern edge of the Eddyville Quadrangle. Well records indicate that the Degonia is present in the subsurface throughout the study area.

The Degonia ranges from about 20 to 64 feet thick in the study area and generally is about 40 feet thick; it is thinnest in the southeastern Eddy-

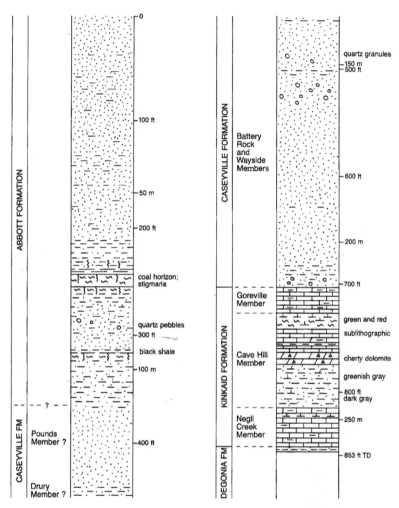

Figure 5 Graphic log of Alcoa test hole, SW NW NE NW, Section 2, T12S, R6E, Eddyville Quadrangle. The borehole is in the Dixon Springs Graben, immediately southeast of the Lusk Creek Fault Zone. (Log is based on core descriptions by Paul E. Potter, William H. Smith, and D. B. Saxby 1956).

ville Quadrangle, but otherwise no consistent thickness trends are apparent. The Degonia-Clore contact is often impossible to determine from well records because the thin limestones of the upper Clore Formation are hard to identify in geophysical logs.

The Degonia thickens to more than 100 feet and becomes dominantly clean, crossbedded, fine-grained sandstone along a southwest-trending belt lying northwest of the study area (Potter et al. 1958, Potter 1963).

Topography The Degonia underlies gently to moderately sloping terrain in its area of outcrop. Discontinuous sandstone ledges occur in areas free of talus from overlying units. Otherwise, outcrops are confined to beds and banks of streams.

11

Lithology The Degonia Formation consists of very fine sandstone, siltstone, shale, claystone, and bedded chert. The lithologic sequence is moderately variable.

Distinctive, bedded chert is found at the base of the Degonia in the Eddyville Quadrangle. Dull white to brownish gray, opaque, and closely fractured, the chert occurs in tabular to irregular beds up to about 1 foot thick. The chert is best exposed in an old road and adjacent streambed near the center of the NE, Section 3, T12S, R6E. This chert also is widespread in the Herod and Karbers Ridge Quadrangles (Baxter and Desborough 1965, Baxter et al. 1967) and in the Waltersburg Quadrangle (Weibel et al. in preparation). The sample study of well 15 (table 3) in the Stonefort Quadrangle indicates chert at the base of the Degonia.

The main part of the Degonia generally consists of a thin shale at the base overlain by sandstone that grades upward to siltstone and silty shale. The sandstone rarely is thicker than about 10 feet. A few well logs, as well as outcrops in the Stonefort Quadrangle, indicate two thin sandstones separated by 10 to 20 feet of shale and siltstone. Shale and siltstone in the Degonia commonly are dark gray, olive, or greenish gray. Sandstone exposed in the Eddyville Quadrangle is bluish green to olive and weathers light brown to yellowish brown; it is very fine grained and forms ripple-marked beds 1 to 4 inches thick. Exposures are found in the beds of Ramsey Branch and Lusk Creek near the road crossings in NW, Section 3, T12S, R6E, and in the road itself east of Lusk Creek. Degonia sandstone exposed in the Stonefort Quadrangle is light gray to light brown, very fine grained, and thin bedded to thick bedded. Sedimentary structures include faint parallel laminations, current ripples, and horizontal tubular burrows.

Red to green variegated claystone or soft shale is a distinctive feature of the uppermost part of the Degonia. This material can be seen in the bank of Lusk Creek just north of the road in NW, Section 3, T12S, R6E; it is also recorded on drillers' logs of many wells in the study area. According to Swann (1963), red shale is found near the top of the Degonia throughout much of southern Illinois.

| SYSTEM | SERIES | | FORMATION | LITHOLOGY | BED, MEMBER |
	Western Europe	N. American midcontinent			
PENNSYLVANIAN	Cantabrian	Desmoinesian	Carbondale		Houchin Creek Coal Mbr
					Survant Coal Mbr
					Colchester Coal Mbr
	Westphalian D				Dekoven Coal Mbr
					Davis Coal Mbr
					sub-Davis sandstone
			Spoon		Carrier Mills Shale Mbr
					Stonefort Ls Mbr
					Wise Ridge Coal Bed
					Mt. Rorah Coal Mbr
					Creal Springs Ls Mbr
					Murphysboro (?) Coal Mbr
					golden sandstone
					Mitchellsville Ls Bed
					New Burnside Coal Bed
					Delwood Coal Bed
	Westphalian C	Atokan			Oldtown Coal Bed
			Abbott		Murray Bluff Ss Mbr
					olive shale
	Westphalian B				middle Abbott sandstone lentils
					lower Abbott sandstone lentils
		Morrowan			Tunnel Hill Coal Bed
	Westphalian A				basal Abbott shale, ss
					Reynoldsburg Coal Bed

The contact of the Degonia with the overlying Kinkaid Limestone is sharp and apparently conformable.

Depositional environment Only limited inferences can be made on depositional environments of the Degonia because of lack of exposures. Body fossils are absent and the few trace fossils observed are not diagnostic. Current ripples indicate directional flow, but the direction of flow differs from one layer to the next; this could reflect either shallow subtidal or nonmarine overbank sedimentation. Crawling traces and fossil plants observed in the Degonia south of the study area suggest a nearshore environment. Variegated red and green claystone like that found at the top of the Degonia may represent a paleosol (Jerzykiewicz and Sweet 1986). The bedded chert at the base of the Degonia may have

12

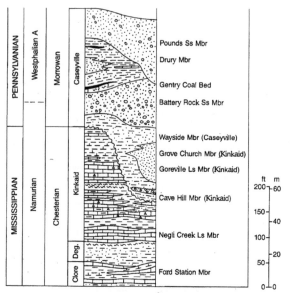

Figure 6 Generalized stratigraphic column of bedrock units exposed in the study area, with correlations to Western European and North American Midcontinental series.

been derived from silicification of calcareous siltstone (Baxter et al. 1967).

Regional studies by Potter et al. (1958) and Potter (1963) suggest that the Degonia represents distal sediments from a distributary system that lay northwest of the study area.

Kinkaid Limestone

Name and correlation
Weller (1920) named the Kinkaid Limestone for Kinkaid Creek in Jackson County, Illinois, approximately 40 miles west of the study area. Swann (1963) revised the definition of the Kinkaid to include three newly named members—the Negli Creek Limestone (basal), the Cave Hill, and the Goreville Limestone. Swann assigned shale above the Goreville Limestone to a new formation, the Grove Church Shale. In this publication we have reclassified the Grove Church as a member of the Kinkaid, because it does not meet the requirement of mappability the North American Stratigraphic Code (1983, article 24d) stipulates for a valid formation. Also, the Grove Church is composed

of interbedded shale and limestone that are lithologically similar to the rest of the Kinkaid. The Kinkaid and its four members (fig. 4) are readily traceable, on the basis of lithologic content in outcrop and subsurface, from their respective type localities to the study area.

Distribution and thickness
The Kinkaid crops out along Ramsey Branch and Lusk Creek in the southeastern part of the Eddyville Quadrangle and along the southwest-flowing stream near the southwest corner of the Stonefort Quadrangle. Two small exposures occur at the crest of the McCormick Anticline in SW, Section 11, T11S, R5E, Stonefort Quadrangle. The Kinkaid is identified in all wells that reach the required depth within the study area. One core description from the Eddyville Quadrangle is available (fig. 5).

The thickness of the Kinkaid varies because of erosion at the base of the overlying Caseyville Formation (figs. 4, 6). The minimum thickness (about 85 ft) and maximum thickness (about 230 ft) of the Kinkaid were both

recorded in well logs in the Creal Springs Quadrangle. In most of the study area 120 to 160 feet of Kinkaid are preserved.

Topography The Kinkaid commonly is obscured by talus on slopes below outcrops of the Caseyville Formation. Where talus is absent the limestones of the Kinkaid form ledges or subtle benches separated by slopes underlain by shale; the best example of this type of topography is found in SE, Section 4, T12S, R6E, Eddyville Quadrangle.

Lithology The Kinkaid Formation consists of limestone and shale and small amounts of siltstone and claystone (figs. 4, 6). The vertical lithologic sequence is consistent throughout the study area, although gradual lateral variation is evident.

The Negli Creek Member (Swann 1963), almost entirely limestone, ranges from 25 to 38 feet thick. The lower part of the Negli Creek generally is a very fine to fine-grained, dark gray to brownish gray limestone; it is argillaceous (although distinct shale partings are rare) and in many places it contains blue-gray to brown chert nodules. Weathered surfaces are rough, and the rock is generally crumbly or rotten. Beds range from a few inches to about 2 feet thick and are irregular or hummocky. Brachiopods, bryozoans, and echinoderm fragments are common, but the most characteristic fossils are large bellerophontid gastropods, *Girvanella* (algal) oncoids, and *Chaetetella*. The gastropods and *Girvanella* occur in other Chesterian limestones, but their occurrence together is definitive for the lower part of the Negli Creek. *Chaetetella*, found in one outcrop of Negli Creek in the study area, has been reported in no other unit in the region (Trace and McGrain 1985). The upper part of the Negli Creek is gray, fine- to coarse-grained, bioclastic limestone dominated by pelmatozoan fragments, fenestrate bryozoans, and brachiopods. Cross-bedding was observed in the upper Negli Creek near the south edge of the Eddyville Quadrangle west of Lusk Creek. The overall lithologic sequence of the Negli Creek in the study area is very similar to that described by Randall (1970) and Buchanan (1985).

The Cave Hill Member (Swann 1963) is divisible into three lithologic

13

units in the study area and elsewhere in southern Illinois. The lower unit is greenish gray shale and siltstone, the middle unit is mainly sublithographic to fine-grained limestone interbedded with shale, and the upper unit is red and green, variegated claystone interbedded with thin limestone. The Cave Hill is as thin as 55 feet where its upper part is missing (the result of pre-Caseyville erosion), but is 75 to 95 feet thick where fully preserved.

The lower shale unit of the Cave Hill thickens to the northeast across the study area; it is 7 to 24 feet thick in the Creal Springs and Stonefort Quadrangles, 20 to 30 feet thick in the southern part of the Eddyville Quadrangle, and 27 to 45 feet thick in the Mitchellsville oil field just north of the Eddyville Quadrangle. Accompanying the increase in thickness is an increase in grain size to the northeast. In the west the lower unit consists of silt-free, dark gray to greenish gray clay-shale, but in the east, silty shale and siltstone are found, particularly in the upper part of the unit. Outcrops are scarce; the description of the lower unit is based mainly on well cuttings and geophysical logs. The northeastward thickening and coarsening correspond to regional trends recognized by Randall (1970).

The middle unit of the Cave Hill becomes thinner and contains more shale to the northeast. The thickness decreases from 50 to 75 feet in the Creal Springs Quadrangle to about 30 feet in the Mitchellsville oil field. The middle unit is at least three-fourths limestone in the Creal Springs, Stonefort, and southern Eddyville Quadrangles but is half shale in the Mitchellsville oil field.

Limestone in the middle unit is sublithographic to fine grained and light to dark gray; most of it weathers to light gray, smooth, rounded surfaces. The limestone is dense and rings when struck with a hammer, unlike limestones of the Negli Creek and Goreville Members, which generally crumble when struck. Some beds are dolomitic and weather yellowish orange. Much of the limestone is argillaceous or silty. Bedded and nodular chert is abundant. Shale interbedded in the middle unit is medium to dark gray or greenish gray, and mostly calcareous. The most complete exposure of the middle unit of the Cave Hill is

along Ramsey Branch in NW NW SE, Section 4, T12S, R6E (Eddyville field station 642).

The middle unit of the Cave Hill contains a rich fauna of brachiopods, bryozoans, molluscs, and echinoderms. Some of the more common fossils are the brachiopods *Spirifer increbescens*, *Composita sp.*, and *Diaphragmus sp.*, the bryozoans *Archimedes sp.*, *Fenestella sp.*, and *Eridopora sp.*, and the gastropods *Platyceras* and dwarf *Bellerophon*. The shells are commonly whole and articulated, but they do not appear to be in life position. Most *Archimedes* axes are intact, although the fronds are detached. Most echinoderm remains, except those of a few *Pentremites* calices, are disarticulated.

The upper unit of the Cave Hill consists of soft, generally noncalcareous, greenish gray and red-green, variegated shale or claystone containing beds of nodular, highly fossiliferous limestone less than 6 inches thick. The bivalves *Myalina* and *Edmondia* are characteristic of this unit. Outcrops are visible along Ramsey Branch in the SE SE, Section 5, T12S, R6E. The upper unit is generally 10 to 20 feet thick throughout the study area.

The Goreville Member of the Kinkaid, like the Negli Creek Member, consists almost entirely of limestone and a few shale partings (Swann 1963). The Goreville is thin or missing in places because of pre-Caseyville erosion, but where fully preserved it is 35 to 45 feet thick. As exposed in the Eddyville Quadrangle and in well cuttings, it is medium to dark gray or brownish gray and fine to coarse grained. Weathered surfaces are rough because of the silicified fossil fragments they contain, and the limestone commonly crumbles when struck. Some parts of the Goreville are highly argillaceous and contain thin layers of olive gray, calcareous shale. Chert nodules and bands up to 24 inches thick are conspicuous in some horizons. The Goreville is composed largely of pelmatazoan fragments; partly articulated crinoid stem segments are common. Bryozoans also are common; very large specimens of *Archimedes* (with axes up to 12 inches long) are diagnostic of the Goreville in the report area.

The Goreville generally becomes coarser grained, and contains more sparry cement and less micrite and clay westward.

The Grove Church Shale Member occurs in places beneath the pre-Caseyville unconformity. The only outcrop found is immediately south of the Eddyville Quadrangle in gullies about 600 feet from the south line, 1,500 feet from the east line, Section 4, T12S, R6E, Waltersburg Quadrangle. A core of the Grove Church was obtained from hole W-1, drilled close to this outcrop (plate 1). The Grove Church also was identified in the logs of six or seven wells in the Creal Springs Quadrangle and one well each in the Eddyville and Stonefort Quadrangles. A thickness of 71 feet of Grove Church shale (close to the known maximum thickness of the Grove Church) was identified in the Jenkins #1 Mohler well (well 21, table 3). In outcrops and well samples the Grove Church consists primarily of olive gray to greenish gray, lightly mottled, partly calcareous shale that is silt free to finely silty. Limestone beds in the lower Grove Church are less than a foot thick, fine to coarse grained, very shaly, and fossiliferous. An interval of limestone up to 18 feet thick is indicated on well logs in the upper Grove Church.

Depositional environment The argillaceous limestone of the lower part of the Negli Creek Member of the Kinkaid probably was deposited in relatively deep water with low to moderate turbulence. Abundant *Girvanella* oncoids suggest gently oscillating currents that rolled fossil grains along the sea bottom. An increase in wave and current energy, probably related to shoaling, is indicated by the coarse, locally crossbedded biosparite of the upper part of the Negli Creek (Buchanan 1985).

The shale and siltstone of the lower part of the Cave Hill Member probably are prodeltaic sediments, from a delta that lay northeast of the study area (Randall 1970). Periodic input of mud from this source continued through accumulation of the middle part of the Cave Hill. Most limestone of the Cave Hill was laid down in gently agitated water, as indicated by slight to moderate disarticulation of delicate fossils.

The red and green mudstone of the upper part of the Cave Hill contains abundant *Myalina* (mussels), suggesting shoaling and possible episodes of subaerial exposure.

14

Within our study area, the Caseyville Formation overlies the Grove Church, Goreville, and Cave Hill Members of the Kinkaid Formation (fig. 6). Total relief on the contact is about 150 feet. Just south of the study area in the northeastern part of the Glendale Quadrangle, the Kinkaid and Degonia Formations are locally absent, and the Caseyville rests directly on strata of the Clore Formation (Devera in preparation).

The Kinkaid Formation in the study area is characterized by a consistent vertical sequence having only minor and gradual lateral variations. This "layer cake" stratigraphy of the Kinkaid contrasts sharply with the lateral lithologic variability of the Caseyville Formation.

Variations in thickness of the Kinkaid clearly are the result of erosion of the upper members and replacement by Caseyville sediments. If depositional continuity between the Kinkaid and the Caseyville exists anywhere in the Illinois Basin, it must occur where the Grove Church Member is most complete (it would be unusual for the contact to be conformable at one or a few sites but unconformable everywhere else in the basin). Only the lower shale of the Grove Church is present at localities about 8 miles west of the report area, where Rexroad and Merrill (1985) collected their fossils. The upper limestone of the Grove Church, observed in boreholes within the study area, is absent at the type locality.

Lateral intertonguing of Caseyville and Grove Church lithologies, which would be expected if the units represented continuous deposition, never has been reported. Rexroad and Merrill asserted that the Caseyville and Grove Church are lithologically similar, but this is not true. The lower part of the Caseyville is dominantly sandstone and siltstone, but these lithologies are absent in the Grove Church. Limestone, greenish gray shale, and calcareous shale, which make up the bulk of the Grove Church, are rare in the Caseyville. Some of the olive, mottled, silty shale of the Grove Church does resemble certain shales in the Caseyville, but the former is fissile or platy and the latter is weakly laminated and commonly interlayered with sandstone.

To summarize, lithologic evidence indicates that depositional continuity between the Grove Church and the Caseyville is highly unlikely. Although the age of the Grove Church has not been completely established, Jennings and Fraunfelter (1986) reported significant floral and faunal breaks between Grove Church and basal Caseyville and suggested that Caseyville conodonts described by Rexroad and Merrill may have been reworked from Mississippian strata.

No clear exposures of the Kinkaid-Caseyville contact were found in the study area. In quarries a short distance to the south, the contact is sharp and gently undulating. Pre-Pennsylvanian downcutting is shown best in the southern part of the Eddyville Quadrangle. Grove Church Shale is preserved in SE, Section 4, T12S, R6E; along the Ramsey Branch near the southwestern corner of the same section, the Caseyville rests on the lower part of the Goreville Member. Pennsylvanian rocks overlie the upper part of the Cave Hill Member in the outcrops in the southwestern Stonefort Quadrangle.

PENNSYLVANIAN SYSTEM

The Pennsylvanian System of the Illinois Basin—particularly the lower part of the system in the study area—is difficult to subdivide into lithostratigraphic units. The system is composed mostly of sandstone, siltstone, and shale, which generally form lenticular bodies that intergrade laterally and vertically. Intercalated with the siliciclastic rocks are thin beds of coal, limestone, and black fissile shale. In the lower part of the Pennsylvanian (Morrowan and Atokan Series) these beds are mostly lenticular, but in Desmoinesian and younger strata they become widely traceable and useful as marker beds.

Most Pennsylvanian formations in the Illinois Basin were defined partly on the basis of observable lithic characteristics (primarily by differences in the composition and texture of sandstones) and partly on the basis of marker beds (especially coals and limestones). Many problems have arisen because lithic features and marker beds that are well developed in type areas cannot be recognized in other parts of the basin. As a result, the formations currently recognized by the Illinois, Indiana, and Kentucky Geological Surveys differ substantially (fig. 7). Recent efforts of the three state surveys to stan-

15

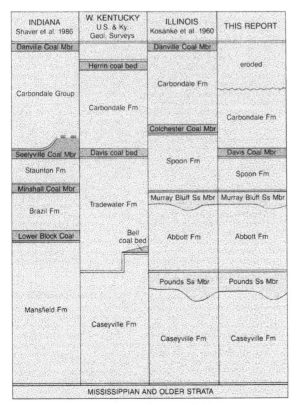

INDIANA Shaver et al. 1986	W. KENTUCKY U.S. & Ky. Geol. Surveys	ILLINOIS Kosanke et al. 1960	THIS REPORT
Danville Coal Mbr		Danville Coal Mbr	
	Herrin coal bed		eroded
Carbondale Group		Carbondale Fm	
	Carbondale Fm		Carbondale Fm
		Colchester Coal Mbr	
Seelyville Coal Mbr	Davis coal bed		Davis Coal Mbr
Staunton Fm		Spoon Fm	
			Spoon Fm
Minshall Coal Mbr		Murray Bluff Ss Mbr	Murray Bluff Ss Mbr
Brazil Fm	Tradewater Fm		
Lower Block Coal	Bell coal bed	Abbott Fm	Abbott Fm
		Pounds Ss Mbr	Pounds Ss Mbr
Mansfield Fm			
	Caseyville Fm		
		Caseyville Fm	Caseyville Fm

MISSISSIPPIAN AND OLDER STRATA

Figure 7 Formational nomenclature of the Pennsylvanian System in the Illinois Basin, as used by the three state geological surveys and in this report. The top of the Caseyville Formation in western Kentucky has been mapped either at the highest occurrence of sandstone containing quartz pebbles or at the base of the Bell coal bed. The Seelyville Coal Member (Staunton Formation of Indiana) commonly is a split seam; the lower bench is equivalent to the Davis Coal and the upper bench correlates with the Dekoven Coal (Jacobson 1987).

dardize stratigraphic terminology have led to the adoption of several common terms for members (Jacobson et al. 1985). A basinwide classification of the Pennsylvanian into formations was proposed during a meeting of the Tri-State Correlation Committee for the Pennsylvanian in April 1989 for eventual adoption by the three surveys.

In this study we adopt, with revisions, the formational nomenclature proposed by Kosanke et al. (1960) and used by Baxter et al. (1963), Baxter and Desborough (1965), Baxter et al. (1967), Nelson and Lumm (1986a,b,c 1990a,b) and Trask and Jacobsen (1990) in geologic mapping of quadrangles east of the present study area. The formations recognized are the Caseyville (oldest), Abbott, Spoon, and Carbondale. Several members and beds have been revised or redefined, others have been deleted, two new formal members have been proposed, and several informal units are introduced.

Caseyville Formation

Name and correlation Owen (1856, p. 48) named the Caseyville conglomerate after the settlement of Caseyville on the Ohio River in Union County, Kentucky. Lee (1916) revised the name to Caseyville Formation and described the type section on the Illinois side of the Ohio River opposite Caseyville. Later, some geologists, such as Weller (1940), referred to the Caseyville as a group. Kosanke et al. (1960) ranked the Caseyville as a formation in Illinois and described a reference section along the Illinois Central Railroad in the south-central part of the Stonefort Quadrangle. The Caseyville has been mapped extensively in western Kentucky and is equivalent to the lower part of the Mansfield Formation in Indiana (fig. 7).

Correlation of the Caseyville Formation into the study area is based on its lithologic similarity to the type Caseyville and on the physical continuity of the Pounds and Battery Rock Sandstone Members from their type localities into these quadrangles.

Distribution and topography The Caseyville Formation crops out extensively in the southern parts of all three quadrangles of this study. It also is exposed in many places along the crest of the McCormick Anticline, which runs westward across the central part of the Eddyville Quadrangle and thence southwestward to the southwest corner of the Stonefort Quadrangle. The Caseyville produces rugged topography; massive sandstones, more than 100 feet thick in places, form bold cliffs, ledges, and steep slopes (fig. 8). A line of cuestas marks the Caseyville outcrop in the southern part of the study area. Intervals of shale, siltstone, and shaly sandstone between the massive Caseyville sandstones erode to gentle slopes and strike-valleys. Talus from the sandstone cliffs commonly covers the nonresistant intervals of the Caseyville.

Thickness The Caseyville Formation varies from about 170 to more than 450 feet thick in the report area. Thickness values were obtained from well records and from composite sections measured on outcrops (table 4). This limited evidence suggests that the Caseyville is generally thicker in synclinal areas than it is on the McCormick and New Burnside Anticlines and adjacent to the Lusk Creek Fault Zone.

16

Figure 8 Typical bluff exposure of Battery Rock Sandstone Member, Caseyville Formation at Cedar Creek, near the southeast corner of the Creal Springs Quadrangle, SW SW, Section 3, T12S, R4E. Note the large talus block of sandstone near the stream (right).

Baxter et al. (1967) reported a thickness of 300 to 500 feet for the Caseyville in the Herod Quadrangle east of the study area. Nelson and Lumm (1986c) reported a thickness of 250 to 300 feet in the Rudement Quadrangle.

Lithology The Caseyville Formation is characterized by white to light gray, relatively clean quartzose sandstone commonly containing well-rounded granules and small pebbles (up to 1 inch) of white quartz. Sandstone in the Caseyville varies from very fine to very coarse grained and from very thin bedded to massive. In most places, one-half to three-fourths of the thickness of the Caseyville is sandstone. The remainder of the formation consists of light to dark gray or brownish gray siltstone; medium to dark gray (rarely black), silty shale; olive gray to brownish gray claystone; thin, discontinuous beds of shaly coal; and localized, dark gray, calcareous shale containing thin beds and nodules of limestone.

The only place in the report area where the Caseyville Formation cannot be differentiated confidently from Chesterian strata is within the small fault slice labeled "Chesterian undifferentiated" (described previously). The upper contact of the Caseyville is more problematic, because lithologic differences between the Caseyville and the overlying Abbott Formation are subtle and gradational. The chief difference is in the character of the sandstones. Caseyville sandstones (both fine and coarse grained) are relatively clean, sparkly, quartz sandstones generally classified as quartz arenites; Abbott and younger Pennsylvanian sandstones contain more mica, feldspar, lithic fragments, and clay matrix and are classified as subgraywackes (Siever and Potter 1956, Potter and Glass 1958). Quartz granules and pebbles are much more abundant in the Caseyville than in younger Pennsylvanian formations. The Pounds Sandstone Member is the youngest sandstone that consistently displays the lithic character of the Caseyville. Therefore, following the practice of most geologists who have worked in southern Illinois and western Kentucky, we have mapped the top of the Caseyville at the top of the Pounds Sandstone. The contact is drawn at the top of the highest ledge of thick-bedded sandstone; any overlying thin-bedded sandstone is assigned to the Abbott. In a small part of the Eddyville Quadrangle where the Pounds Sandstone was not identified, the top of the Caseyville was placed at the highest occurrence of clean quartzose sandstone.

Four previously named members of the Caseyville Formation are recognized in the study area. These are (in ascending order) the Wayside, the Battery Rock Sandstone, the Drury, and the Pounds Sandstone. The Gentry Coal Bed, which occurs within the Drury Member, has also been identified.

Table 4 Thickness of the Caseyville Formation in the study area

Location			Thickness (ft)
Eddyville Quadrangle			
Outcrop section	SE	34-11S-6E	200
Water well, Eddyville village			257
Oil-test holes		12-11S-5E	235-310
Milo Ditterline #1 Hart	SE SE NW	14-11S-6E	300
Roy Pledger #1 Gibson	NW NE NW	13-11S-5E	405
Stonefort Quadrangle			
Outcrop sections near southwest corner of quadrangle			200-250
Garnier et al. #1 Peeples	NE NE NE	30-11S-5E	260
Pierce #1 Camden	NW NW NW	35-10S-5E	313
Ohio Oil Co. #1 Bynum	NW NW NE	35-10S-5E	465?
Creal Springs Quadrangle			
Mitchel and Stanonis #1	SE NE SE	23-11S-3E	170
Bessie Mohler #1	SW SW NW	24-11S-3E	200
Biver #1	NW SW NE	24-11S-3E	235
Charles Wrightsman #1	SE SW NW	17-11S-4E	230
Parson #1	NE NW NE	16-11S-4E	290
Evans #1	SE SE NW	10-11S-4E	220
Horn #1	SE NW NW	5-11S-4E	310

17

Wayside Member Lamar (1925) applied the name, Wayside Sandstone and Shale Member, to basal Caseyville strata below his Lick Creek (now Battery Rock) Sandstone Member in the Carbondale 15-minute Quadrangle, about 8 miles west of our study area. Kosanke et al. (1960) shortened the name to Wayside Sandstone Member. Weller (1940) applied the name Lusk Formation (Caseyville Group) to the same interval of strata in the Waltersburg Quadrangle immediately south of the Eddyville Quadrangle. Kosanke et al. (1960) revised the name to Lusk Shale Member, without giving reasons for the change or clearly outlining the difference between the Lusk and the Wayside. Mapping in progress demonstrates that the type Wayside and the type Lusk are lithologically similar units of interbedded shale and sandstone and are stratigraphically equivalent. The name Lusk is therefore abandoned in favor of Wayside, which has priority. Also, the lithic term "sandstone" is omitted, because the Wayside contains large proportions of shale and siltstone.

The Wayside Member is generally nonresistant, but in places it contains cliff-forming sandstones. Much of the Wayside is concealed by talus and colluvium on slopes below the Battery Rock Sandstone. The Wayside is not known to be absent anywhere in the study area, but in some areas exposures are too poor to permit mapping this member.

The Wayside Member consists of variable proportions of shale, siltstone, and thin- to medium-bedded sandstone, along with minor conglomerate and lenticular bodies of massive sandstone. Coal has been reported in the Wayside in other areas, but none was found in our three quadrangles. Shale in the Wayside is medium to dark gray or brownish gray, silty, finely micaceous, and noncalcareous. Thin intervals of black carbonaceous shale also occur. Wayside siltstone, light to medium gray or olive gray, commonly occurs as lenses or laminae in shale. Thin-bedded sandstones of the Wayside are white to light gray or brown, very fine-grained, quartzose, and very well indurated. Bed thickness is quite variable; most thin-bedded intervals include some layers that are several feet thick. The most common sedimentary structures in thin-bedded sandstones are current and interference ripples and small load casts. The Wayside contains a greater abundance and variety of load structures than any other unit in the study area. Trace fossils in the Wayside include simple, branching, or curving horizontal burrows, crawling traces and feeding traces, and escape burrows inclined to bedding. The only organic remains found in thin-bedded Wayside in the study area are poorly preserved plant fragments.

Good exposures of thin-bedded Wayside strata are along the west-flowing stream in the NW SE, Section 2, T12S, R4E, in the southwestern part of the Stonefort Quadrangle. The upper part of the Wayside is exposed in the railroad cut south of the tunnel in Section 31, T11S, R5E, Stonefort Quadrangle (fig. 9). In the Eddyville Quadrangle the Wayside crops out along the Ramsey Branch and adjacent gullies.

A minor but distinctive lithology found only in the Wayside is conglomerate consisting of clasts, up to several inches in diameter, of sandstone, ironstone, and chert in a matrix of highly ferruginous sandstone and claystone. Such conglomerate occurs as beds or lenses up to a few feet thick interlayered with thin-bedded sandstone and shale. Outcrops were observed near the center of Section 35, T11S, R6E, Eddyville Quadrangle, and in the stream bed in SW SE NE, Section 2, T12S, R4E, Stonefort Quadrangle.

Thick-bedded to massive cliff-forming sandstones up to 40 feet thick occur in the Wayside Member in the Stonefort and Eddyville Quadrangles. These sandstones are notably lenticular in comparison with the widely traceable Pounds and Battery Rock Sandstones higher in the Caseyville. Massive Wayside sandstones are generally fine grained and lack the quartz pebbles that are abundant in the Battery Rock and Pounds. Crossbedding is seldom as conspicuous in Wayside sandstones as in younger Caseyville sandstones. Fossils include rare coalified logs, along with silicified corals and crinoid fragments reworked from Mississippian limestones. Wayside sandstone is thickest south of the study area in the northwestern part of the Waltersburg Quadrangle and northeastern part of the Glendale Quadrangle, where a deep paleochannel was eroded into Mississippian strata (Devera, in preparation). Cliffs of Wayside sandstone up to 40 feet high are present along Hayes Creek at the southern edge of the Eddyville Quadrangle and south of the stream near the center of Section 2, T12S, R4E in the Stonefort Quadrangle.

The upper contact of the Wayside Member is well exposed only in the railroad cut south of the tunnel in Section 31, T11S, R5E, Stonefort Quadrangle. Here, as in outcrops in adjacent quadrangles, the contact of the Wayside with the overlying Battery Rock Sandstone Member is sharp and appears to be erosional.

The Wayside varies in thickness from less than 30 feet to more than 150 feet. The Wayside is at least 150 feet thick in W 1/2, Section 1, and in E 1/2, Section 2, T12S, R4E, in the southwestern part of the Stonefort Quadrangle. Within a mile southwest of this area the Wayside thins to 35 to 50 feet. In the Eddyville Quadrangle the Wayside thickens from less than 30 feet along the upper part of Ramsey Branch to at least 80 feet along the lower part of the same stream; it is as thick as 115 feet in the subsurface in the Eddyville Quadrangle. We do not have enough data to determine whether or not these variations are related to irregularities on the sub-Pennsylvanian surface.

Battery Rock Sandstone Member The Battery Rock Sandstone Member was named by Cox (1875) for Battery Rock, a bluff along the Ohio River in eastern Hardin County, Illinois. Mapping in the fluorspar district (Baxter et al. 1963, Baxter and Desborough 1965, Baxter et al. 1967) established the physical continuity of the type Battery Rock with sandstone in our study area. The Battery Rock has been identified as far west as Jackson County, Illinois (Desborough 1961, Sonnefield 1981).

The Battery Rock is a resistant, cliff-forming sandstone (fig. 8); it is well exposed along the drainages of Lusk Creek and Hayes Creek in the Eddyville Quadrangle, Bay Creek and Little Bay Creek in the Stonefort Quadrangle, and Cedar Creek in the Creal Springs Quadrangle. It also is exposed in several places along the McCormick Anticline. The sandstone is exposed in railroad cuts along the Illinois Central tracks in the southern part of the Stonefort Quadrangle.

18

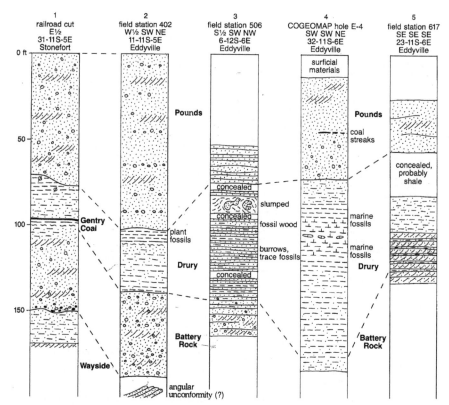

Figure 9 Measured sections of the Caseyville Formation.

The Battery Rock is a thick-bedded to massive, conglomeratic, quartzose sandstone. Nearly all outcrops of Battery Rock contain abundant quartz pebbles 1/4 to 1 inch in diameter. Lenses of quartz-pebble conglomerate are common, especially near the base of the unit. Pebbles are larger and more numerous in the Battery Rock than in the younger Pounds Sandstone Member.

Nearly white when fresh, Battery Rock sandstone weathers to light or medium gray. Weathered surfaces are smooth and rounded; the sandstone has a sugary texture. Large-scale planar and trough crossbedding is common, and ripple marks occur locally. The only fossils are transported logs of *Lepidodendron* and *Calamites*. Vertical burrows believed

to be *Skolithos* were observed near the railroad just south of the Stonefort Quadrangle (NW NW, Section 17, T12S, R4E).

The contact of the Battery Rock with thin-bedded sandstone and shale of the overlying Drury Member is abrupt in most places. In the Illinois Central railroad cut (SE, Section 31, T11S, R4E) the Battery Rock is directly overlain by the underclay of the Gentry Coal Bed. Near the southern boundary of the Stonefort Quadrangle in the railroad cut, the upper part of the Battery Rock grades laterally to siltstone and thinly laminated, shaly sandstone. Gradational Battery Rock-Drury contacts are exposed in a gully alongside a trail west of Eddyville (fig. 9, third column), near the center, W 1/2, Section 6, T12S,

R6E and in an east-flowing small tributary of Little Lusk Creek just north of the SE corner, Section 23, T11S, R6E, also in the Eddyville Quadrangle.

The Battery Rock is 30 to 60 feet thick in most places; it may reach 100 feet or more along Bear Branch and Lusk Creek in the Eddyville Quadrangle, but at these locations it cannot be easily differentiated from massive sandstones of the Wayside Member. The Battery Rock becomes thin and lenticular east of Lusk Creek and has not been mapped. It is probably 75 to 100 feet thick along Burden Creek and Caney Branch, near the crest of the McCormick Anticline in the east-central part of the Stonefort Quadrangle; well records here indicate as much as 145 feet of massive

19

sandstone in the lower Caseyville, but part of this may be Wayside sandstone.

Drury Member Lamar (1925) named the Drury Shale and Sandstone Member of the Pottsville Formation for Drury Creek, about 20 miles west of the present study area, in the Carbondale 15-minute Quadrangle. Lamar's Drury was an interval of nonresistant shale and sandstone between his Lick Creek (Battery Rock) and Makanda (Pounds) Sandstones. Kosanke et al. (1960) recognized the Drury as a member of the Caseyville Formation and redefined it as the Drury Shale Member. We are herein changing the name to Drury Member because in many places half or more of the interval consists of sandstone.

The continuity of the Drury from the area mapped by Lamar to our study area has been verified by mapping (in progress) in the intervening Goreville Quadrangle, west of the Creal Springs Quadrangle (Jacobson, in preparation).

Mapping in our three quadrangles shows the Drury overlying the Battery Rock Sandstone in nearly all the area where the Battery Rock is present. One exception is in E 1/2, Section 1, T12S, R4E, Stonefort Quadrangle, where the Drury has been eroded and replaced by Pounds Sandstone. The same situation may exist in the southwestern part of the Creal Springs Quadrangle. In the southwestern part of the Eddyville Quadrangle, thick talus deposits that conceal the Drury interval have been mapped.

Where strata are horizontal or gently dipping, the Drury erodes to a topographic bench between the Pounds and Battery Rock escarpments. Along the McCormick Anticline where the beds are tilted, the Drury underlies strike-valleys. In both situations most of the Drury is hidden by talus and colluvium derived from the Pounds Sandstone. Outcrops of the Drury thus generally are fragmentary except along a few streams where downcutting is especially rapid.

The Drury, like the Wayside, consists principally of variable proportions of light to dark gray clay-shale and silty shale, light gray, thinly laminated siltstone, and light gray, very fine-grained, thin- to medium-bedded, shaly sandstone. Pebbly

sandstone and conglomerate are rare. Local lenses of crossbedded sandstone up to about 15 feet thick produce small ledges. One such sandstone lens is well exposed on the south side of Cedar Creek in SE NE NE, Section 5, T12S, R4E, Creal Springs Quadrangle. Coal and carbonaceous shale occurring in the Drury will be discussed in a separate section. Dark gray to black calcareous shale containing thin bands or nodules of limestone has been found in the southern part of the Eddyville Quadrangle.

The most complete exposure of the Drury Member is along a north-flowing tributary of Caney Branch (fig. 9, column 2). At the top of the unit, about 8 feet of dark gray clay-shale, containing plant fossils, overlies 2 feet of soft, gray claystone. Below the claystone is 27 feet of medium-dark gray, silty shale containing thin laminae and lenses of light gray siltstone and sandstone. Both the upper and lower contacts of the Drury are sharp at this locality. A nearly complete exposure of the Drury occurs along a small gully beside a trail just west of Eddyville (fig. 9, column 3). Here the member is 65 to 70 feet thick, and both upper and lower contacts are gradational. The Drury at this site consists of thin- to medium-bedded sandstone interbedded with siltstone and shale. Burrows and other trace fossils are abundant in the lower part of the Drury and casts of fossil wood occur near the middle part.

The lower part of the Drury is well exposed along a ravine west of Little Lusk Creek (fig. 9, column 5). Near the mouth of the ravine about 30 feet of sandstone is present. This sandstone, transitional from Battery Rock to Drury, occurs in beds a few inches to 4 feet thick and contains scattered quartz pebbles. The sandstone beds have planar crossbedding in which the foreset beds consistently dip southwest and pass into ripple-marked bottomset beds. The lower unit grades upward to about 20 feet of ripple-marked siltstone containing laminae of fine-grained sandstone. At the top of the Drury is an interval about 20 feet thick that contains float and small, slumped exposures of dark gray shale.

The upper part of the Drury and the overlying Pounds Sandstone are exposed at Peter Cave near the southwest corner of the Eddyville Quadrangle (field station 508). The

Drury here consists of 25 to 30 feet of yellowish gray, very fine-grained, silty sandstone having herringbone crossbedding; its base is concealed. The Pounds-Drury contact is erosional, showing sharp truncation of bedding and 12 to 15 feet of vertical relief. Small sandstone dikes from the Pounds penetrate the uppermost Drury near the back of the "cave" (actually a rock shelter).

Dark gray to black shale containing abundant marine fossils was encountered at four sites in the Drury Member of the south-central part of the Eddyville Quadrangle. The sites are (1) gullies beside an abandoned road near the center of the west line of Section 33, T11S, R6E (field station 519); (2) a cutbank on a tributary of the Bear Branch, near the center of S 1/2, Section 32, same township (field station 518); (3) gullies beside a dirt road, NW SE NW, Section 5, T12S, R6E (field station 515); and (4) a drill core from borehole E-4, near center Section 32, T11S, R6E. In the outcrop a maximum of 8 feet of shale was exposed in the upper part of the Drury. The drill core (fig. 9, column 4, and plate 1) includes 82 feet of medium dark gray to black, partly calcareous shale containing thin bands or nodules of argillaceous limestone. The shale grades upward to sandstone of the Pounds and downward to sandstone of the Battery Rock. Fossils in the shale include the goniatites *Axinolobus*, *Gastrioceras*, and *Wiedeyoceras (?)*, nautiloids, bivalves, gastropods, brachiopods, conodonts, and palynomorphs (Devera et al. 1987).

The only other reported occurrence of calcareous marine strata in the Drury Member is from a bluff of the Ohio River near the abandoned community of Sellers Landing, eastern Hardin County, Illinois (Wanless 1939, 1956, Baxter et al. 1963). The limestone, which occurs near the base of the Drury Member, is thought to be slightly older than the Gentry Coal Bed (Wanless 1956). The marine shale in the Eddyville Quadrangle lies near the top of the Drury and is palynologically younger than the Gentry Coal; therefore, the marine shale of the Eddyville Quadrangle probably is younger than the Sellers Limestone. The Drury is thin to absent where it was cut out and replaced by the Pounds Sandstone and as thick as 120 feet in the southeastern part of the Creal Springs

20

Quadrangle. In general, the Drury thins where the Pounds thickens, and vice versa. The Pounds-Drury contact tends to be sharp and erosional where the Drury is thin and gradational where the Drury is thick (fig. 9; plate 1).

Gentry Coal Bed A coal bed that crops out near Battery Rock on the Ohio River in Hardin County, Illinois, was named the Battery Rock Coal by Owen (1856). Kosanke et al. (1960) renamed the coal the Gentry Coal Member to avoid duplicating the name used for the Battery Rock Sandstone. We are in this report changing the rank of the Gentry Coal from a member to a bed (1) because a member cannot contain another member (North American Stratigraphic Code, 1983, Article 26), and (2) to establish a hierarchy in which local or lenticular coals like the Gentry are ranked as beds, and widely continuous coals like the Davis and Herrin are ranked as members. In so doing, we depart from previous ISGS practice, in which all coals were ranked formally as members (e.g., Davis Coal Member) and from current USGS and Kentucky Survey policy, in which all coals are ranked informally as beds (e.g., Bell coal bed, with *coal bed* in lower case). The distinction between a coal bed and a coal member, in our usage, will be decided on a case-by-case basis. In general, we are ranking as members those coals that are traceable across several counties, and ranking as beds the less traceable coals.

The Gentry Coal has been identified in parts of Hardin County, Illinois (Baxter et al. 1963, Baxter and Desborough 1965), and Crittenden and Livingston Counties, Kentucky (Amos 1965, 1966). This coal is not physically traceable from these sites to the study area, where it is probably lenticular. The coal we are calling Gentry is in the same stratigraphic position as the type Gentry and is similar in palynological content.

The Gentry Coal is exposed at several places in the southwestern part of the Stonefort Quadrangle, but its crop line has not been plotted on the geologic map because of space constraints. The best exposures are in railroad cuts in the southwestern part of the Stonefort Quadrangle (fig. 9, column 1). The coal was also observed near the head of Jackson Hollow in NE NW NW, Section 6,

T12S, R5E (Stonefort field station 781) and in a small gully in NE NW SW, Section 36, T11S, R4E (field station 775). At all these localities the coal is 14 to 24 inches thick and is very shaly, except for the uppermost 4 to 6 inches, which is brightly banded. The coal, found about 3 to 10 feet above the top of the Battery Rock Sandstone, is separated from the Battery Rock by soft shale and claystone; it is overlain by silty shale and siltstone in the natural exposures and by thin- to medium-bedded sandstone in the railroad cut.

Coal identified palynologically as Gentry was sampled from wastepiles of two collapsed drift mines alongside a northeast-trending stream in SW NW NE, Section 7, T11S, R6E, Eddyville Quadrangle (field station 387). Coal also is recorded in drillers' logs of two oil-test holes in NE, Section 12, T11S, R5E, about 3/4 mile west of the drift mines. Both logs indicate that the coal occurs at the base of the Drury Member. A weathered outcrop of coal was found near the base of the Drury in a stream in SE NE SW, Section 12. Black shale containing plant fossils was found at the position of the Gentry Coal in NE SW SE, Section 33, T11S, R6E, Eddyville Quadrangle.

Unnamed coal in Drury Member
Thin, shaly coal about 5 feet below the top of the Drury was sampled from a small ravine in SW NW NE, Section 3, T12S, R5E (Stonefort field station 6). This bed, 18 to 24 inches thick, is palynologically distinct from the Gentry Coal. Shale that contains plant fossils and may either overlie or be laterally equivalent to a coal has been observed in upper Drury strata at three sites in the study area. One site is at Eddyville field station 402 (fig. 9, col. 2). A second is near the stream junction in SW SE NW, Section 36, T11S, R4E (Stonefort field station 758). The third is a ravine in SW NW SE, Section 4, T12S, R4E, Creal Springs Quadrangle (Nelson's field station 9).

Pounds Sandstone Member The Pounds Formation in the Caseyville Group was introduced by Weller (1940); its type locality is Pounds Hollow in southern Gallatin County, Illinois. Kosanke et al. (1960) revised the name to Pounds Sandstone Member, Caseyville Formation, and defined the top of the Pounds as the

top of the Caseyville Formation in Illinois. This definition has been used by numerous mappers working in southern Illinois and adjacent parts of western Kentucky (Amos 1965, 1966, Baxter et al. 1963, Baxter and Desborough 1965, Baxter et al. 1967, Desborough 1961, Nelson and Lumm 1986a,b,c) and is used in this report.

The Pounds Sandstone is a cliff-forming unit in much of southern Illinois. Cliffs of this sandstone, which are relatively continuous from the Ohio River to the Mississippi, are popularly called the Pounds Escarpment. Many scenic cliffs of Pounds, and deep gorges, waterfalls, and rockshelters are found in the study area; these include Double Branch Hole, Jackson Hole, and Peter Cave in the Eddyville Quadrangle, and Bell Smith Springs, Burden Falls, Cedar Falls, and Jackson Hollow in the Stonefort Quadrangle. Large areas of bare rock are common at the tops of cliffs and knobs capped by Pounds. Bold hogbacks of Pounds Sandstone mark the north and northwest flanks of the McCormick Anticline, especially in the area northeast of Burden Falls. In contrast, the Pounds forms gentle slopes on some of the drainage divides (for example, in the village of Eddyville).

The Pounds is similar lithologically to the Battery Rock Sandstone, but generally has fewer and smaller quartz pebbles than the latter. Some Pounds outcrops contain no pebbles. In most places quartz granules and pebbles less than 1/2 inch in diameter are scattered throughout the Pounds or concentrated along certain bedding planes. Conglomerate is rare. Bedding and sedimentary structures of the Pounds resemble those of the Battery Rock. Occasional fossil logs of *Calamites* and *Lepidodendron* have been observed. Bivalve resting traces (*Pelecypodichnus*) were found in float believed to be from the Pounds along Bear Branch in the Eddyville Quadrangle.

The Pounds is thin to absent in some places and more than 120 feet in others. It is at least 100 feet thick in Jackson Hollow in the southwestern part of the Stonefort Quadrangle, where the Drury Member underlies the Pounds. Westward, where the Drury Member apparently was eroded, the Pounds rests directly on the Battery Rock Sandstone. Continuous exposures of sandstone are present from the 510-foot to the

700-foot contour in SE, Section 1, T12S, R4E. More than 120 feet of sandstone, probably all Pounds, crops out as cliffs east of the stream in E 1/2, Section 3, T12S, R3E, Creal Springs Quadrangle. The Pounds is 60 to 105 feet thick on the McCormick Anticline northeast of Burden Falls. Between Bell Smith Springs and Eddyville the sandstone ranges from 40 to 75 feet thick. Eastward from the center of Section 32, T11S, R6E, Eddyville Quadrangle, the Pounds thins abruptly (within 1/2 mile) from about 50 feet to 5 feet. The Pounds was not recognized in the two small exposures of Caseyville on the McCormick Anticline in Section 2, T11S, R6E, Eddyville Quadrangle.

The lower contact of the Pounds was discussed previously. The upper boundary of the Pounds, and thus of the Caseyville Formation, is mapped at the top of the highest cliff or ledge of sandstone. The cliff tops correspond with the highest occurence of thick-bedded to massive sandstone. Fine-grained, thin-bedded, flaggy sandstone, commonly overlying the cliff- and ledge-forming sandstone, is classified as Abbott Formation.

In the small area in the east-central part of the Eddyville Quadrangle where the Pounds is absent, the top of the Caseyville was mapped at the highest occurence of sandstone having typical Caseyville lithology. Caseyville sandstone in this area is white to light gray, nearly pure quartz containing small quartz pebbles; it varies from thin to thick bedded but does not produce a continuous ledge.

Age of the Caseyville Formation

Evidence for the age of the Caseyville Formation is based on fossil pollen and spores from coal and shale and on goniatites from the marine shale of the Drury Member in the Eddyville Quadrangle. Palynological data indicate that the Caseyville is of Morrowan age, according to the standard series terminology in the Pennsylvanian System of the North American Midcontinent region (fig. 6). The Morrowan corresponds to the Namurian C and Westphalian A stages of the Upper Carboniferous Series in western Europe. The goniatites confirm the findings of palynology. The genus *Axinolobus*, identified in the Drury, is associated with late Morrowan or Westphalian A strata (Devera et al. 1987).

Depositional environment Previous studies of lower Pennsylvanian sandstones in the Illinois Basin (Potter and Siever 1956, Potter and Glass 1958, Potter 1963) established a dominant southwesterly paleocurrent direction. This direction conforms to the trend of sub-Pennsylvanian paleovalleys (Bristol and Howard 1971). Most geologists, including those cited above and Ethridge et al. (1975) and Koeninger and Mansfield (1979), have interpreted the Caseyville as primarily the product of fluvial and deltaic sedimentation. These authors also attribute the quartz-arenitic character of Caseyville sandstones to derivation from sedimentary and low-grade metamorphic rocks in the northern Appalachians or southeastern Canadian Shield.

The transport of such a quantity of sand and gravel (particularly as much as that seen in the Battery Rock and Pounds Sandstones) such a distance from the source area would seem to require a relatively steep gradient and a wet climate. Orogenic uplift of the source area would provide the increased gradient. Early Pennsylvanian tectonism in the northern Appalachians has been well documented in West Virginia (Englund et al. 1986), eastern Pennsylvania (Wood et al. 1986), New England (Skehan et al. 1986), and Nova Scotia (Bell 1938, Rust et al. 1987). The paleoclimate of the Morrowan Epoch evidently was wet (Phillips and Peppers 1984, Cecil et al. 1985), although not as wet as that of the late Desmoinesian Epoch.

Although they are not continuous deposits, the Pounds and Battery Rock Sandstones in the study area are more like "blanket" sandstones than like the linear or "shoestring" sands typically associated with fluvial or distributary channels. These sandstones have some characteristics of braided channel systems (Brown et al. 1973). Another possibility is that Pounds and Battery Rock sands were reworked by marine processes. Detailed sedimentological work will be required to test these hypotheses.

The presence of goniatites in the Drury Member in the southern part of the Eddyville Quadrangle establishes marine influence there. Low diversity of fauna, large numbers of juvenile goniatites, and lateral facies relations suggest a drowned distributary channel. High content of organic carbon (1.7%) and replacement of

fossils by marcasite are evidence of dysaerobic conditions (Devera et al. 1987). The herringbone crossbedding in the Drury Member at Peter Cave indicates tidal currents. Nonmarine conditions elsewhere in the Drury are indicated by the presence of coal, rooted underclay, and shale containing unbroken plant fossils.

Marine sedimentation in the Wayside Member has been documented to the west (Rexroad and Merrill 1985, Jennings and Fraunfelter 1986) and to the south (Devera, in preparation) of the present study area.

Abbott Formation

Name and definition The Abbott Formation was proposed by Kosanke et al. (1960) and named for Abbott, an abandoned station on the Illinois Central Railroad near the center of the Stonefort Quadrangle. (The name is spelled "Abbot" on the current 7 1/2-minute topographic map, but was spelled "Abbott" on the older 15-minute Harrisburg Quadrangle map). The type section consists of railroad cuts north and south of Abbot (fig. 10). The Abbott Formation was defined there as comprising strata from the top of the Caseyville Formation to the top of the Murray Bluff Sandstone Member. The type locality of the Murray Bluff Sandstone is at Murray Bluff, near the northwest corner of the Eddyville Quadrangle about 4 miles east of the Abbott type section. The Murray Bluff Sandstone is present (although incorrectly identified by Kosanke et al. 1960) in the Abbott type section (fig. 10).

Kosanke et al.'s definition of the upper boundary of the Abbott Formation has caused difficulties for geologists attempting to map this formation. Kosanke et al. defined overlying Spoon Formation in west-central Illinois more than 200 miles from the Abbott type area. Abbott and Spoon lithologies grade into one another in both areas. The base of the Spoon was defined as the top of the Bernadotte Sandstone Member, which Kosanke et al. believed equivalent to the Murray Bluff Sandstone. That equivalence has never been verified; Kosanke et al. did not provide a means of differentiating Abbott and Spoon Formations in areas where the Bernadotte/Murray Bluff is absent or unidentifiable. In spite of these difficulties, we have elected to retain the Abbott and

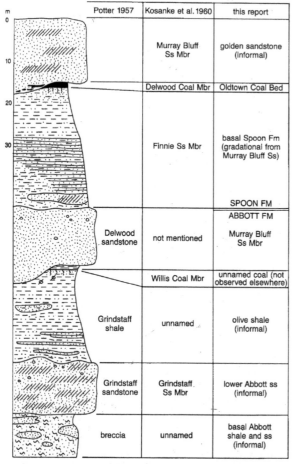

	Potter 1957	Kosanke et al. 1960	this report
		Murray Bluff Ss Mbr	golden sandstone (informal)
		Delwood Coal Mbr	Oldtown Coal Bed
		Finnie Ss Mbr	basal Spoon Fm (gradational from Murray Bluff Ss)
			SPOON FM
			ABBOTT FM
	Delwood sandstone	not mentioned	Murray Bluff Ss Mbr
		Willis Coal Mbr	unnamed coal (not observed elsewhere)
	Grindstaff shale	unnamed	olive shale (informal)
	Grindstaff sandstone	Grindstaff Ss Mbr	lower Abbott ss (informal)
	breccia	unnamed	basal Abbott shale and ss (informal)

Figure 10   Type section of Abbott Formation, Illinois Central Railroad cuts, Stonefort Quadrangle, with nomenclature of Potter (1957), Kosanke et al. (1960), and this report.

Spoon Formations in the present study. The interval of rock between the Murray Bluff Sandstone and the Pounds Sandstone is noticeably sandier and produces a more rugged topography than the interval above the Murray Bluff, particularly in the eastern part of the study area. With practice, geologists can generally distinguish sandstones from above and below the Murray Bluff in outcrop and hand specimen. The Murray Bluff itself is traceable through most of the report area, although its character varies considerably from place to place.

We mapped the Abbott-Spoon contact at the top of the highest ledge of thick-bedded Murray Bluff Sandstone. Overlying thin-bedded sandstone, present in many places, was assigned to the Spoon Formation. In some areas where the Murray Bluff is absent or unrecognized, the Abbott-Spoon contact has been projected and is indicated by a dashed line on the geologic maps.

Distribution and topography   The Abbott Formation covers most of the Eddyville Quadrangle except (1) the southeastern and extreme southwestern parts and a narrow strip along the McCormick Anticline where it is eroded; and (2) parts of the northern quarter of the quadrangle where the Spoon Formation overlaps it. In the Stonefort Quadrangle the Abbott crops out southeast of the McCormick Anticline and on the structurally higher parts of the New Burnside Anticline. It also is exposed along all the ravines between the two anticlines and is thinly covered by the Spoon Formation on the intervening uplands. The Abbott is at the surface in most of the southern half of the Creal Springs Quadrangle and in the valleys of Wagon and Larkin Creeks; it is eroded along the large streams in the extreme south and is covered by thin, irregular patches of Spoon Formation on uplands in the southeast.

The Abbott produces a topography of broad, gently rolling uplands deeply incised by ravines. The break in slope between valleys and uplands generally forms at the top of a sandstone. Ledges and local cliffs of sandstone line the valley walls, but no long escarpments like those of the Pounds Sandstone are developed. Alternating sandstones and shaly intervals produce hogbacks and strike valleys, respectively, on the flanks of anticlines.

Thickness   Few reliable thickness measurements of the Abbott Formation are available because of the absence of continuous sections and the difficulty of picking the contacts in subsurface logs. Composite sections must be used with caution because individual units vary considerably in thickness. The Abbott is estimated to be 350 to 400 feet thick in the Eddyville Quadrangle; it may be thicker in the Dixon Springs Graben at the southeast corner of the quadrangle, where the upper part of the Abbott is eroded. The Alcoa core (fig. 5) indicates 350 feet as a minimum thickness.

The Abbott thins abruptly from about 350 to 250 feet in the eastern half of the Stonefort Quadrangle. Approximately 200 feet is exposed in

23

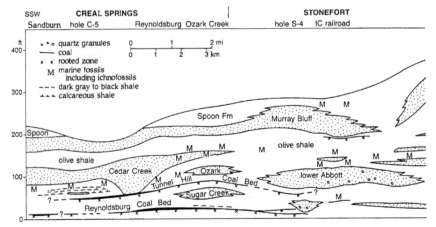

Figure 11 Generalized stratigraphic cross section of Abbott Formation from the northeastern part of the Eddyville Quadrangle to the southwestern part of the Creal Springs Quadrangle. Datum, top of Caseyville. Stippled pattern represents a ledge-forming sandstone; unpatterned areas represent shale, siltstone, and thin-bedded sandstone.

the railroad cuts of the type section along the north-south centerline of the quadrangle (fig. 10), and an additional 50 feet or so is concealed at the base. Test hole S-4 (plate 1) cored the entire Abbott Formation. The lower contact is fairly definite; the upper part of the Abbott grades from thin- to medium-bedded sandstone (having shale breaks) to crossbedded sandstone. If the top of the crossbedded sandstone is defined as the top of the Abbott, the unit is about 250 feet thick. Estimated to be 220 to 300 feet thick in the Creal Springs Quadrangle, the Abbott continues to thin westward in the Goreville Quadrangle (Jacobson, in preparation).

Lithology The Abbott Formation, lithologically transitional between the Caseyville and Spoon Formations, has greater lateral variability than either. It consists almost entirely of siliciclastics and a few lenses of coal and rare limestone, but the proportions and vertical distribution of sandstone, siltstone, and shale vary widely from one place to another.

Sandstones of the Abbott are very light gray to medium gray when fresh and typically weather brown. Grain size ranges from very fine to coarse. Widely scattered quartz granules and small pebbles up to about 1/4 inch are present, but abundant large pebbles like those found in the Caseyville

do not occur. Conglomerates of shale pebbles or molds of pebbles—and less commonly, coal and ironstone clasts—in sandstone matrix are fairly widespread. Petrologic maturity of sandstones decreases upward through the Abbott. Lower Abbott sandstones closely resemble Caseyville sandstones except that they generally lack quartz pebbles and have slightly more mica, dark grains, and interstitial clay. Upward in the Abbott, mica grains, feldspar, and rock fragments become increasingly common, and clay matrix occurs instead of silica cement. Iron oxide is prominent, especially in the middle part of the Abbott. These sandstones weather dark brown and commonly have iron-oxide boxwork, fracture-fillings, and Liesegang banding. Because of their clay matrix and weak cementation, Abbott sandstones are generally more friable and less resistant to erosion than Caseyville sandstones.

Abbott sandstones are not as widely traceable as the Pounds and Battery Rock Members of the Caseyville; some are traceable only a few tens to hundreds of yards. Larger sandstone bodies locally are more than 100 feet thick but in most places are less than 50 feet. Relatively tabular Abbott sandstones typically range from less than 10 to about 40 feet thick.

Siltstones and shales of the Abbott and Caseyville are generally indistinguishable. Bioturbation is more widespread in Abbott shale and siltstone than in the Caseyville. Certain intervals in the Abbott that are pervasively bioturbated serve as local markers within the study area. Dark gray to black calcareous shale containing marine fossils and thin layers or nodules of limestone are locally present.

The Abbott Formation is not completely classified into members in this publication. Several previously named sandstone members and coal beds or members have proved to be lenticular and are difficult or impossible to identify away from their type areas. Only the Reynoldsburg Coal Bed and the Murray Bluff Sandstone Member, which have type localities within the study area, can be identified with assurance. All other units within the Abbott shown on the geologic maps and described herein are named informally. Correlations of some of these informal units around the study area are tenuous, and in some areas one or more of the units cannot be identified. The following units have been mapped and are discussed in this report:

Murray Bluff Sandstone Member (youngest)
Olive shale (informal)

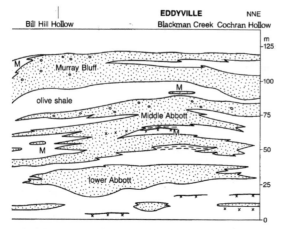

Middle Abbott sandstone lentils, including Cedar Creek sandstone lentil (informal)
Lower Abbott sandstone lentils, including Ozark sandstone lentil (informal)
Basal Abbott shale and sandstone, including Sugar Creek sandstone lentil (informal)
Tunnel Hill Coal Bed
Reynoldsburg Coal Bed

The geometric relationships of these units across the study area are illustrated in figure 11.

Basal Abbott shale and sandstone The basal unit of the Abbott Formation in our study area consists of shale, siltstone, thin-bedded sandstone, and local lenses of thick-bedded sandstone (figs. 5, 12). The Reynoldsburg Coal Bed occurs near the base and the Tunnel Hill Coal near the top; both coals are discontinuous.

The dominant lithologies of the basal unit are shale, siltstone, and thin-bedded sandstone. Medium gray to dark gray clay-shale and silty shale commonly is thinly interlaminated with light gray siltstone or sandstone. The thin-bedded sandstones, which closely resemble those of the Caseyville, consist of white to very light gray, very fine, slightly impure, well-indurated quartz sand. Bedding is typically flaggy and may display ripple marks and trace fossils. Flaggy sandstone, especially common at the base of the Abbott,

grades upward from the Pounds Sandstone.

The basal unit has been mapped only in an area along the McCormick Anticline in the Eddyville Quadrangle, where overlying lower Abbott sandstones are relatively continuous. In areas where lower Abbott sandstone is absent, the basal unit cannot be separated lithologically from younger strata; such areas are mapped simply as Abbott Formation, undifferentiated. For discussion purposes, the basal unit includes that part of the Abbott Formation beneath lower Abbott Sandstone lentils in the Stonefort Quadrangle and beneath the Tunnel Hill coal and Ozark sandstone in the Creal Springs Quadrangle.

The total thickness of the basal unit of the Abbott is about 50 to 80 feet in the Creal Springs Quadrangle and 50 to 150 feet in the Eddyville and Stonefort Quadrangles.

The basal unit erodes to gentle slopes and strike valleys and commonly is covered with talus and colluvium derived from younger sandstones. Thick-bedded sandstones produce ledges or small cliffs in places; otherwise the interval is exposed only in ravines and artificial excavations.

Among the few relatively complete exposures of the basal unit are several in cutbanks and gullies along tributaries of Blackman Creek, in NW SW, Section 2, and E 1/2 SE, Section 3, T11S, R6E, Eddyville Quadrangle (fig. 11). Several coarsening-upward

cycles of strata were observed; each begins with dark gray clay-shale at the base and coarsens upward through silty shale and siltstone to sandstone. Two of the cycles are topped off by gannister (extremely hard, fine-grained, orange-weathering sandstone with a lumpy appearance) that is thoroughly penetrated by *Stigmaria*. Thin coal was observed overlying one of the gannisters. The entire interval is about 150 feet thick, and individual cycles are 30 to 70 feet thick.

Some lenticular sandstones within the basal member are thick enough to have been mapped. One such sandstone occurs along Bear Branch (SW NW, Section 32, T11S, R6E, Eddyville Quadrangle). This sandstone, 12 to 15 feet thick at most, lies about 25 feet above the base of the Abbott; it is light gray, very fine grained, and thickly bedded. Sandstone of similar lithology, mapped as Sugar Creek sandstone lentil, crops out about 70 feet above the base of the Abbott along Ozark Creek, SW, Section 27, T11S, R4E, Creal Springs Quadrangle. Up to 40 feet of fine-grained, nearly massive sandstone, also mapped as Sugar Creek sandstone lentil, is exposed in ledges and railroad cuts along Sugar Creek in SW NE SW, Section 17, T11S, R4E. The Tunnel Hill Coal directly overlies this sandstone. A persistent sandstone not on the map forms a small ledge 30 to 40 feet above the base of the Abbott along the northwest flank of the McCormick Anticline between McCormick village and Burden Falls in the Stonefort Quadrangle.

The railroad cut north of the tunnel in Section 19, T11S, R5E, Stonefort Quadrangle, reveals unusual lithologies in the basal unit of the Abbott (fig. 10). The basal unit here consists of numerous lenses of sandstone interbedded with shale-pebble conglomerate. The sandstone in the lenses is relatively clean, fine-grained, and quartzose. Sandstone lenses 15 feet thick and as long as 100 feet terminate abruptly, locally interfingering with the conglomerate. The conglomerate consists of gray shale clasts in carbonaceous siltstone and sandstone. In most places the bedding is subhorizontal, but locally the rock has undergone small-scale folding and faulting. Up to about 30 feet of sandstone/conglomerate is sharply and irregularly overlain by cross-bedded to massive sandstone. Potter

25

(1957) described the exposure in detail and called the sandstone/conglomerate a breccia. He interpreted it as the product of subaqueous slumping or gravity flow triggered by tectonic movements along the McCormick Anticline, south of the railroad cut.

Fossils found in basal Abbott strata include plant remains and trace fossils. The ichnofossil *Conostichus sp.* was collected from this interval along a tributary of Lusk Creek in NW NE NW, Section 28, T11S, R4E, Eddyville Quadrangle.

Reynoldsburg Coal Bed Weller (1940) named the Reynoldsburg Coal for the village of Reynoldsburg in the southwestern part of the Creal Springs Quadrangle. No type section was designated, but reference was made to mines in Section 32, T11S, R4E, just west of the village. Kosanke et al. (1960) ranked the Reynoldsburg Coal as a member. We are revising it to a bed because of its limited extent and discontinuous nature. We have mapped the Reynoldsburg Coal in S 1/2, Sections 31 and 32, and in parts of adjacent sections, along Cedar Creek. The coal is known from a few outcrops and numerous abandoned drift mines and prospect pits. This coal was also extracted from two commercial strip mines. The Herod Mining Company had a pit in SE, Section 32, and Energy Exploration Company's Ozark Mine was in SW NE SE, Section 31. According to annual coal reports of the Illinois Department of Mines and Minerals, the Herod mine was operated from May through July, 1965, and produced 1,938 tons of coal; the Ozark Mine operated from July through December, 1978, and turned out 10,417 tons of coal.

The Reynoldsburg Coal was 27 to 31 inches thick at the Ozark mine and 15 to 17 inches at the Herod Mine (ISGS field notes, open files). Field notes on other mines, prospect pits, and outcrops indicate that the coal reaches a maximum thickness of about 36 inches in the Reynoldsburg area. The coal thins rapidly west of the Cedar Creek drainage; it is only 3 inches thick in the core of borehole C-5 in SE SE, Section 36, T11S, R3E (plate 1). The coal was not seen in the southwestern Creal Springs Quadrangle, but it occurs farther west in the Goreville Quadrangle (Jacobson, in preparation).

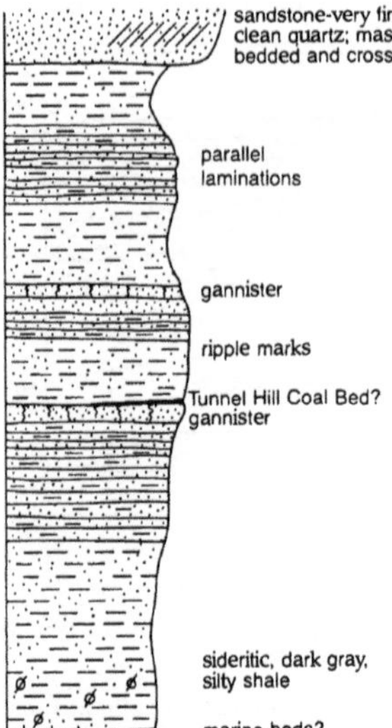

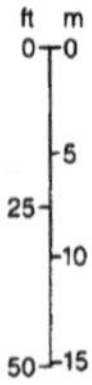

Figure 12 Measured composite section of basal Abbott shale and sandstone sequence as exposed in Blackman Creek, SE SE NE, Section 3 to SW NW SW, Section 2, T11S, R6E, Eddysville Quadrangle.

East of Reynoldsburg, the Reynoldsburg Coal crops out along Ozark Creek in S 1/2, Section 27, T11S, R4E, and has been traced along the north side of Cedar Creek to NW, Section 36, same township, in Stonefort Quadrangle. On Ozark Creek, the dull, hard, finely laminated coal has been described as cannel or oil shale. Barrett (1922) described the deposit and ran distillation tests on the samples. A maximum yield of 48.8 gallons of oil per ton was recovered. Barrett reported that local farmers mined the material and burned it in their cookstoves and fireplaces; no attempts at commercial mining and distillation have been made.

A sample of the Reynoldsburg cannel from Ozark Creek was submitted to James C. Hower of the Kentucky Energy Cabinet Laboratory for petrographic analysis. Hower (written communication 1988) described the sample as a torbanite (boghead coal) and gave its composition as follows: bituminite, 63.2 percent; alginite (*Botryococcus*), 28.8 percent; inertinite, 3.6 percent; exinite, 2.1 percent; vitrinite, 1.2 percent; and resinite, 0.1 percent.

Farther east, a thin coal was found at five sites along the drainage of Hunting Branch, in Sections 27 and 28, T11S, R5E, Stonefort Quadrangle. This coal is 50 to 60 feet above the top of the Pounds Sandstone, whereas the type Reynoldsburg is 20 to 30 feet above the Pounds. The coal along Hunting Branch ranges in thickness from less than an inch to about 10 inches; it is dull and shaly and in places consists of discontinuous lenses or stringers of coal in shale or sandstone. No rooting was observed below the coal. The palynological assemblage is abnormal and differs from that of type Reynoldsburg Coal. Abundant herbaceous lycopod spores and *Laevigatosporites* in two samples suggest that the coal formed on a flood plain or in a similar, relatively dry setting rather than in a peat swamp. A third sample contained acritarchs (marine microfossils possibly related to planktonic algae), which are rare in coal. These findings, plus the field setting, indicate that the coal on Hunting Branch was formed from transported plant material rather than from an in situ peat deposit. Whether it is equivalent to the Reynoldsburg is problematic.

A shaly coal about 5 inches thick occurs about 10 feet above the Pounds Sandstone on a tributary of Bear Branch, SW SE NW, Section 32, T11S, R6E. Sandstone below this coal is thoroughly rooted. Coal palynologically similar to Reynoldsburg coal was sampled from underwater at the stream junction in the SE NE SE, Section 5, same township.

Tunnel Hill Coal Bed A coal that crops out 40 to 70 feet above the Reynoldsburg has been mined at several places in the Creal Springs Quadrangle. It is herein named the Tunnel Hill Coal Bed, for the nearby community of Tunnel Hill. This coal is palynologically equivalent to the Bell (1B) coal bed of western Kentucky, but it is lenticular and probably not continuous with the Bell Coal. A surface mine was operated in the Tunnel Hill Coal by the Holly Mining Company in NW NW, Section 6, T12S, R4E. According to records of the Illinois Department of Mines and Minerals, the mine operated only in August, 1970, and produced 2,542 tons. Judging by the size of the area mined, the tonnage figure appears too small; another company may have worked the mine before or shortly after Holly Mining Company. Several open cuts expose the overburden strata and remnants of the coal.

Several caved drift mines and prospect pits for the Tunnel Hill Coal were found along the valley of Cedar Creek near the Holly strip mine. Evidence of digging also exists around the head of the south-trending ravine in Section 1, T12S, R3E. A collapsed adit containing fragments of coal was noted in a steep side ravine of Sugar Creek, SE NW SW, Section 17, T11S, R4E. Thin coal sampled at two localities in the Eddyville Quadrangle is palynologically similar to the Tunnel Hill (and Bell) Coals. The sites are a small ravine near the NW corner, Section 33, T11S, R6E, and a steep gully in SE NW NE, Section 23, same township.

The Tunnel Hill Coal is 12 to 14 inches thick where it is exposed in the Holly strip mine. Three test holes encountered the coal (plate 1), which was 12 inches thick in hole C-6, 20 inches thick in hole C-3, and 28 inches thick in hole C-5. The coal from all three cores is dull and shaly.

Lower Abbott sandstone lentils
Prominent cliff- and ledge-forming sandstones are widespread above the basal unit in the lower part of the Abbott Formation. These sandstones are as thick as 100 feet, and their bases lie 50 to 150 feet above the base of the Abbott Formation. They do not form a single tabular body of sandstone, but rather a series of lenses at similar stratigraphic position (fig. 11). The name Grindstaff Sandstone

has been applied widely to sandstone in the lower part of the Abbott. The type Grindstaff Sandstone is in southwestern Gallatin County, Illinois (Butts 1925). Baxter and Desborough (1965) and Baxter et al. (1967) mapped the Grindstaff in the fluorspar district east of our study area. Potter (1957, 1963) and Kosanke et al. (1960) called the lowest sandstone in the Abbott type section Grindstaff. However, no sandstone in our study area can be linked with the Grindstaff Sandstone of the fluorspar district, and the lower Abbott sandstones here are too discontinuous to warrant formal naming. We therefore refer to these sandstones informally as lentils within the lower Abbott.

The lower Abbott sandstones are consistently white to light gray when fresh and light to medium gray when weathered. Iron-enriched zones and Liesegang bands are common. Most of the sand is fine grained, but small granules are scattered throughout. Quartz pebbles as large as 1/4-inch in diameter are uncommon. The rock is massive to thick bedded; crossbedding and slumped or contorted laminations are common. Toward the lateral edges of lentils the sandstone becomes more thin bedded and intertongue with enclosing shales.

The most extensive lower Abbott sandstone body crops out along the McCormick Anticline from Section 2, T11S, R6E, Eddyville Quadrangle, westward to NE SE, Section 10, T11S, R5E, Stonefort Quadrangle. This sandstone, 50 to 60 feet thick in most of this area, is especially prominent on the north flank of the anticline, where it forms large hogbacks. Westward, this sandstone gradually thins and pinches out. Eastward, it splits into two benches, and both become thin bedded and grade laterally into shale east of Cochran Hollow (fig. 11). The northward extent is unknown because the sandstone dips below drainage. On the south it intergrades with flaggy, thin-bedded sandstone.

Another lentil of lower Abbott sandstone is well exposed on the west side of Burden Creek, along Ogden Branch, and near the top of the ridge between Burden Falls and McCormick in the Stonefort Quadrangle. This lentil, 40 to 100 feet thick near Ogden Branch, thins rapidly to the southeast. Sandstone in the bed of Allen Branch west of McCormick

and in Katy Reid Hollow south and southeast of Abbot probably is part of the same sand body. In this area quartz granules and small pebbles are fairly common in the sandstone. The sandstone is well exposed and about 40 feet thick in the railroad cut in Section 19, T11S, R5E, which is part of the Abbott type section (fig. 10). Its lower contact is sharp and appears to be erosional, and the upper contact of the exposure also is sharp; in places the upper surface of the sandstone undulates strongly. A strongly rolling upper surface along Allen Branch explains the intermittent exposure of the sandstone in the stream bed. The overlying shale contains small folds, which probably reflect compactional adjustments to the rolling surface of the sandstone.

Several small lenses of lower Abbott sandstone have been mapped in the southeastern Stonefort Quadrangle, along Bay Creek near Watkins Ford and along Hunting Branch and its tributaries. Three lenses of sandstone occur south of Hill Branch in Sections 32 and 33, T11S, R5E. Cliffs up to 40 feet high are present just north of the center of Section 32. Conglomerate of pea-sized quartz granules is exposed on the small knob in NW NE, Section 32. The stratigraphic position of the sandstones in this area ranges from 20 to more than 100 feet above the base of the Abbott.

A lower Abbott sandstone mapped as the Ozark sandstone lentil crops out near the head of Ozark Creek in Section 27, T11S, R4E and on the north sides of Cedar and Sugar Creeks in the Creal Springs Quadrangle. This sandstone also has been mapped on the New Burnside Anticline from Section 17, T11S, R4E, Creal Springs Quadrangle, to SE, Section 3, same township, western Stonefort Quadrangle.

The lower Abbott sandstone lentils intergrade laterally with thin-bedded sandstone, siltstone, and shale (fig. 12). The sandstone typically occurs in flaggy beds less than 4 inches thick, but some beds are as thick as 3 feet (fig. 13). Current ripples and oscillation ripples are common. Gray siltstone and silty shale are interbedded with the sandstone. Good exposures of these lithologies are found along Bear Branch and in adjacent roadcuts beside Rt. 145 north of Eddyville, and in the vicinity of Watkins Ford. All of these thin-

Figure 13 Typical planar-bedded sandstone of the lower Abbott Formation in a tributary of Bay Creek in the southern part of the Eddyville Quadrangle (W 1/2 SE SE, Section 26, T11S, R5E).

bedded strata have been mapped as Abbott Formation, undifferentiated, or as lower Abbott on the Creal Springs Quadrangle.

Middle Abbott sandstone lentils
The middle part of the Abbott Formation contains many mappable lenticular bodies of sandstone that underlie and interfinger laterally with fine-grained clastic strata (fig. 12). The fine-grained strata, designated as the "olive shale" unit, are discussed and mapped separately from the sandstone lentils.

The sandstones described in this section represent part of what Weller (1940) called the Delwood Formation (in the Tradewater Group). The Delwood Formation was named for the village of Delwood in the Eddyville Quadrangles, but no type section was designated. Kosanke et al. (1960) recognized three sandstone members in the Abbott Formation of southern Illinois: the Grindstaff, Finnie, and Murray Bluff Sandstones. The Finnie Sandstone, in the middle Abbott, was named by Owen (1856) in southwestern Union County, Kentucky. Kehn (1974) stated that the Finnie is discontinuous within the Dekoven (Kentucky) Quadrangle, which contains the type section. Palynological studies of coal by Peppers and Popp (1979) determined that the type Finnie is approximately equivalent to the Grindstaff Sandstone. Correlation of the Finnie into southern Illinois (Kosanke et al. 1960, Baxter et al. 1963, 1967, Baxter and Desborough 1965), therefore is

questionable. Middle Abbott sandstones in our report area are too lenticular to warrant formal names.

Middle Abbott sandstones are thick and widespread in the Eddyville and easternmost Stonefort Quadrangles; they crop out along the New Burnside and McCormick Anticlines and in the drainages of Bay, Lusk, and Little Lusk Creeks. In most of the central and western Stonefort Quadrangle no mappable sandstones occur in the middle Abbott. A widespread but generally thin sandstone informally named the Cedar Creek lentil can be traced through most of the area along and southeast of the New Burnside Anticline in the Creal Springs Quadrangle.

Middle Abbott Sandstones form prominent cliffs in only a few places, notably along Cochran Hollow and Spring Valley Creek in the northeastern part of the Eddyville Quadrangle. More commonly, a series of sandstone ledges separated by talus-covered benches (presumably shale and siltstone) is found in an interval 90 to 135 feet thick. Individual sandstone bodies have been mapped where the map scale permits (e.g., Cedar Creek sandstone lentils on the Creal Springs Quadrangle map), but elsewhere the entire interval of sandstone and shale is mapped as middle Abbott sandstone, as in the drainage area of Bay Creek and along the McCormick Anticline from Cochran Hollow westward. Individual cliff-and ledge-forming sandstones have been distinguished in the area south of the anticline and east of Rt. 145.

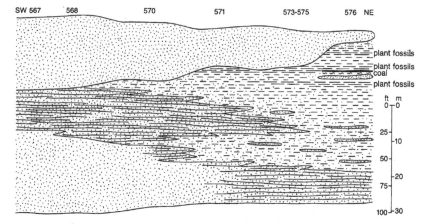

SW 567   568   570   571   573-575   576 NE

Figure 14 Lateral facies change in the middle and lower part of the Abbott Formation in Dixon Springs Graben, southeast of Little Lusk Creek, in the extreme southeastern part of the Eddyville Quadrangle. Lateral distance is approximately 1.2 miles (2 km). Numbers along the top of the figure designate field stations.

quartzose sandstone in tabular, thin to thick beds. Upward, the sandstone becomes medium to coarse grained and poorly sorted and contains some quartz granules. Mica, lithic fragments, and interstitial clay content increase upward. Iron oxide is prevalent; the sandstones weather dark brown, and Liesegang bands and heavy surface crusts of iron oxide are prominent on cliffs and ledges. Sedimentary structures of the upper part of the middle Abbott include planar and trough crossbedding, scour-and-fill, and slumped bedding.

Light gray, thinly laminated siltstone and shaly sandstone, gray to black shale, and thin lenses of coal are interbedded with middle Abbott sandstones in the central part of the Eddyville Quadrangle. Dark gray to black carbonaceous shale was found near the base of the interval in several places; a good outcrop is near the fork of Blackman Creek in SE NE NE, Section 3, T11S, R6E (Eddyville field station 108). A 2-inch coal bed crops out in the stream bed in NW SW SW of the same section (field station 183). Several coal outcrops and collapsed adits were observed in the stream southwest of Zimmer Cemetary near the boundary of Sections 9 and 16 in the same township. Stream cuts reveal 12 to 14 inches of coal overlain by sandstone and underlain by rooted underclay. Palynological analysis

indicates approximate correlation with the Tarter or Willis Coal, which occurs in the middle part of the Abbott. A coal seam 6 inches thick was encountered at a depth of 130 feet in borehole E-3 (plate 1), SE NE NE, Section 9. Several thin, carbonaceous shales or coaly stringers were logged in borehole E-2, near the center of the north line, Section 8, same township (plate 1).

Middle Abbott sandstones interfinger westward with the olive shale unit in the eastern Stonefort Quadrangle. The middle part of the Abbott thins markedly as the sandstones pinch out. The interval from the top of lower Abbott sandstones to the base of the Murray Bluff Sandstone decreases from about 200 to 230 feet in the Eddyville Quadrangle to 80 to 100 feet in the central Stonefort Quadrangle.

The sandstone mapped as Cedar Creek sandstone lentil in the Creal Springs Quadrangle is equivalent to the lower part of the olive shale to the east; it has been traced over a large area that includes the New Burnside Anticline and the drainage areas of Cedar, Sugar, and Little Cache Creeks south of the anticline. Although thin and poorly exposed in most places, the sandstone marks the abrupt change in slope at the tops of the steep-walled valleys and ravines of these three creeks.

The Cedar Creek sandstone generally is light gray to buff when fresh, weathering to brown. Iron oxide is abundant. Grain size is fine to medium, rarely coarse; granules are absent. Mica, lithic fragments and clay matrix are noticeable. Bedding varies from medium to very thick and is generally irregular. Where the sandstone is thick, planar crossbedding is conspicuous.

The Cedar Creek sandstone and underlying shaly strata in the Creal Springs Quadrangle contain abundant ichnofossils and local body fossils, both of which indicate marine sedimentation. The sandstone is strongly bioturbated in many places, particularly in the railroad cut at Tunnel Hill and the highwall of the abandoned strip mine in NW NW, Section 6, T12S, R4E. The ichnofossils *Conostichus broadheadi* (fig. 15), *Rhizocorallium*, and *Pelecypodichnus* were identified here and in nearby outcrops. Calcareous, dark gray to black shale containing thin bands and nodules of argillaceous to sandy, very fossiliferous limestone occur in the same interval. Outcrops are common in the Sugar Creek drainage area southeast of the New Burnside Anticline, and fossiliferous strata were cored in boreholes C-3 and C-6 (plate 1). The following fossils were identified from limestone in the Creal Springs Quadrangle:

29

Brachiopods: *Spiriferellina campestris; Punctospirifer morrowensis; Derbyoides nebraskensis; Antiquatonia sp.; Linoproductus planiventralis; Composita sp.*

Crinoids: *Diphuicrinus c.f. croneisi* (2 calyces); *Sepacocrinus* **plates**
Conodonts: *Idiognathoides ouachitensis; I. noduliferous; I. sinuosis; Idioproniodus conjunctus; Hideodus minutus*

Trilobites: *c.f. Paladin morrowensis*
Bryozoans, bivalves, ostracodes, and nautiloid cephalopods.

In addition, the inarticulate brachiopods *Lingula* and *Orbiculoidea*, productid brachiopods, and the trace fossil *Zoophycus* are found in black shale associated with the limestone. The conodonts suggest late Morrowan to early Atokan age (Jacobson 1983).

The Cedar Creek sandstone is so poorly exposed at most places that its thickness is difficult to estimate. Ravine cuts and artificial exposures typically show 15 to 20 feet of sandstone. The sandstone thickens abruptly to 40 feet or more in some places, forming cliffs. The elongate, north-trending bodies of this sandstone display planar crossbedding with foresets consistently dipping southward. One such body lies along the east side of the large ravine in SW, Section 20, T11S, R4E. Another is near the head of Cache Creek in SW, Section 27, T11S, R3E. The sandstone varies greatly in thickness and interfingers with shale. The exposure in the ravine near the SE corner of Section 36, T11S, R3E is almost entirely sandstone; 2,000 feet southward a thin sandstone bed overlies siltstone, shale, and thin-bedded, shaly sandstone at the same elevation. North of Cedar Creek in SW, Section 31, T11S, R4E, sandstone has been cut down almost to the Tunnel Hill Coal, but fine-grained strata overlie the coal at the abandoned Holly surface mine south of the creek.

Olive shale An interval consisting mostly of shale and siltstone and some thin beds of sandstone partly overlies and partly is laterally equivalent to middle Abbott sandstones described previously (fig. 11). Several geologists working in the early 1900s (unpublished field notes, open files, ISGS) referred to this interval as the "olive shale," and we have adopted this term as an informal mapping unit. The origin of the name is

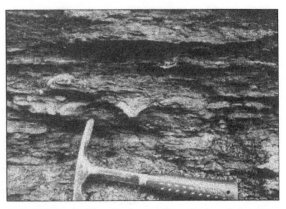

Figure 15 The trace fossil *Conostichus*, interpreted as a resting-trace of a sea anemone and an indicator of marine deposition, in shaly siltstone of middle Abbott Formation, near the south end of the railroad cut south of Tunnel Hill, Creal Springs Quadrangle (SE NE SE, Section 35, T11S, R3E).

unknown; the color of the shale is not noticeably olive.

In the Eddyville and eastern Stonefort Quadrangles the olive shale forms slopes and strike-valleys between the middle Abbott sandstones and the Murray Bluff. Exposures, generally confined to small gullies, are small and fragmentary. The exposed rocks consist mainly of medium to dark gray, silty shale interlaminated with light- to medium-gray siltstone and very fine sandstone. In a narrow northeast-trending ravine in SE NW, Section 35, T10S, R5E, Stonefort Quadrangle, about 25 feet of siltstone in the lower part of the olive member is exposed. The siltstone, ripple marked and shaly in the lower part, becomes intensely bioturbated and less shaly upward. Exposures of part of the olive shale also can be found in small ravines just east of Murray Bluff in the Eddyville Quadrangle.

Numerous ichnofossils are found in the olive shale throughout the study area. The most common form is *Planolites*, a simple horizontal, straight to slightly curved, tubular burrow 1/4- to 3/8-inch in diameter. These burrows, typically filled with sand, weather out of their shale or siltstone matrix. *Planolites* was observed in all three quadrangles in more than a dozen localities; it is found in the upper part of the olive shale in strata approximately equiva-

lent to the Cedar Creek sandstone. Other common ichnofossils are *Pelecypodichnus* and *Cochlichnus*; these types are especially common in the lower part of the olive shale along Ogden Branch and in drainage ditches beside the railroad tracks south of Abbott (Stonefort Quadrangle). A specimen of *Conostichus* was found in shale in the ditch east of the railroad.

Calcareous sandstone and impure, sandy limestone containing fragmentary marine fossils were observed in the olive shale at field station 86 in the Eddyville Quadrangle. The outcrop lies along the base of the bluff on the west side of Blackman Creek, just south of the mouth of a small ravine that runs east from the center of Section 34, T10S, R6E. Old field notes (ISGS, open files) reported other unconfirmed exposures of limestone nearby. Fossils in the limestone consist of broken brachiopod shells and crinoid fragments. The limestone underlies and grades into 10 to 12 feet of sandstone that forms a small ledge along the west side of the valley. Solution channels are developed in the lower part of the limestone. Limestone and dark gray shale containing gastropods, cephalopods, and other marine fossils were found in the upper part of the olive shale in the cores of boreholes C-2 and C-8 (plate 1).

In the Creal Springs Quadrangle, the Cedar Creek sandstone and the shaly interval between the Cedar Creek and Murray Bluff Sandstones are equivalent to the olive shale farther east. The olive shale was not designated on the Creal Springs geologic map. The interval between the Cedar Creek and Murray Bluff Sandstones is 20 to 70 feet thick and is largely covered with colluvium. Test holes C-2, C-6, and C-8 cored this interval (plate 1).

Murray Bluff Sandstone Member Weller (1940) named the Murray Bluff Sandstone and assigned it to the Macedonia Formation in the Tradewater Group. Murray Bluff is the name of a hill near the northwest corner of the Eddyville Quadrangle, where the sandstone is especially prominent. Kosanke et al. (1960) reclassified the Murray Bluff as the uppermost member of the Abbott Formation in southern Illinois.

The Murray Bluff Sandstone has been mapped in all three quadrangles northwest of the McCormick Anticline but has not been positively identified southeast of the anticline. Westward, it continues into the Goreville Quadrangle (Jacobson, in preparation) and appears to die out in the eastern Lick Creek Quadrangle (Weibel and Nelson, in preparation). Desborough (1961) did not identify the Murray Bluff in the Pomona Quadrangle, Jackson County. East of our study area, Baxter et al. (1965) and Baxter and Desborough (1967) traced the Murray Bluff through the Herod and Karbers Ridge Quadrangles. Nelson and Lumm (1984, 1986a,b,c) were unable to identify the Murray Bluff north of the Fluorspar District.

The thickest known occurrence of Murray Bluff is at the type locality; here it is 115 feet thick, and the top is eroded. Along Battle Ford Creek in the Eddyville Quadrangle and along the Little Saline River from Sand Hill eastward in the Stonefort Quadrangle, the Murray Bluff is 100 feet or thicker. The sandstone becomes shaly southward and thins to less than 50 feet on the flank of the McCormick Anticline. The Murray Bluff grades laterally to shale, siltstone, and thin-bedded sandstone in Sections 6, 8, and 16, T11S, R5E, Stonefort Quadrangle. The Spoon-Abbott contact here is mapped at the projected position of the top of the sandstone. Farther west the sandstone reappears and becomes as thick as 80 feet along Clifty Creek and the Little Saline River in the northwestern part of the Stonefort Quadrangle. Much shale and siltstone are interbedded with the sandstone in this area. In the southwestern Stonefort Quadrangle and throughout the Creal Springs Quadrangles, the Murray Bluff ranges from less than 10 feet to about 40 feet thick.

The Murray Bluff splits into two sandstone benches separated by a parting of shale or siltstone on the north flank of the New Burnside Anticline in the Eddyville Quadrangle (fig. 11). The parting appears about 1 mile east of the type locality and thickens eastward to about 30 feet. The upper sandstone bench is generally 20 to 25 feet thick; the lower bench is 50 to 80 feet thick. Good exposures are present in ravines in NE SE Section 28, T10S, R6E, and in roadcuts along Rt. 145 in Section 27, same township. The core of borehole E-1 also indicates two benches of Murray Bluff (plate 1). The splitting is not indicated on the geologic map.

Sandstone similar in lithology and stratigraphic position to the Murray Bluff occurs south of the McCormick Anticline in the Eddyville Quadrangle. Because of its uncertain correlation with the Murray Bluff, this sandstone is designated as upper Abbott. It has been mapped on hilltops near Oak and in the area southeast of Bay Creek and west of Rt. 145. Outcrops are small, and most are confined to stream cuts, roadcuts, and discontinuous ledges. The most accessible exposure is a roadcut along Rt. 145 in NW SW NW, Section 29, T11S, R6E.

The topography of the Murray Bluff is less consistent than that of older Pennsylvanian sandstones. Where it is thick and massive, the sandstone commonly forms cliffs (fig. 16) separated by steep slopes having float and small outcrops. The distribution of outcrops could give the impression that the Murray Bluff consists of a series of lenses; however, fresh exposures in gullies commonly reveal thick sections of massive or thick-bedded sandstone in areas where no cliffs or ledges are present. The lithology of the Murray Bluff in such areas does not differ noticeably from that of the cliff-forming Murray Bluff. Differential resistance to erosion may be related to differ-

31

Figure 16 Murray Bluff Sandstone at its maximum development just east of the type locality in NE NW NE, Section 35, T10S, R5E, in the northwesternmost part of the Eddyville Quadrangle. The sandstone dips 15° to 20° northwest on the flank of the New Burnside Anticline.

ences in cementation of the sandstone. On cliffs and ledges the Murray Bluff tends to be heavily impregnated with iron compounds and has conspicuous Liesegang banding (fig. 17). This sandstone is case hardened: the impregnated crust protects the more friable rock beneath. Elsewhere, notably on Sand Hill in the Stonefort Quadrangle, the Murray Bluff disintegrates to sand. Where it is thin, the Murray Bluff erodes to occasional ledges on moderately sloping, float-covered hillsides.

Lithologically, the Murray Bluff is less mature than older Pennsylvanian sandstones. Quartz is the main constituent, but mica, feldspar, rock fragments, and interstitial clay occur in greater quantities than in older sandstones. When fresh, the sandstone is light gray to medium gray, or brown; it weathers golden brown to dark brown and commonly displays Liesegang bands. Where the Murray Bluff is thick, it is generally fine to coarse grained and contains scattered small, rounded, quartz granules, especially in the upper part of the unit. Thinner Murray Bluff normally is very fine to fine grained and lacks granules; its bedding differs according to thickness of the unit. Thick Murray Bluff is thick bedded to massive. Planar foreset beds in sets as thick as 6 feet are conspicuous in many outcrops. As the Murray Bluff thins, its bedding also thins, and shale or siltstone interbeds are seen. Crossbedding

becomes less prominent and ripple marks more numerous. In the western Stonefort and Creal Springs Quadrangles the Murray Bluff is generally finer grained, more shaly, and less massive than it is to the east. Quartz granules have not been noted, but lenses of shale- and coal-pebble conglomerate are found in the east-central Creal Springs Quadrangle. Gray, silty shale and siltstone commonly interfinger with the sandstone. Ripple marks, load casts, and bioturbation can be seen in the shale and siltstone.

The contact of the Murray Bluff with the underlying olive shale unit is sharp and is probably erosional in most places. The best exposures of a clearly erosional lower contact are in railroad cuts in NW, Section 18, T11S, R5E, Stonefort Quadrangle. A sandstone body that is probably a paleochannel or paleoslump is exposed along the ravine near the center of NW, Section 4, T11S, R5E (Stonefort field station 608) and is indicated on the geologic map. This sandstone is approximately 1,400 feet long and 200 feet wide where exposed and is incised 30 to 40 feet into the olive shale. It is linear (or nearly so) and strikes N20°E. The western contact resembles a fault surface dipping 60° to 70° east; however, no indication of tectonic faulting was found. The eastern margin is concealed. Bedding in the sandstone is horizontal, or nearly so, and the adjacent shale is poorly exposed.

The upper boundary of the Murray Bluff generally is gradational through an interval of thin- to medium-bedded, shaly sandstone to the overlying siltstone and shale of the Spoon Formation (plate 1). The contact is mapped at the top of thick-bedded sandstone.

Age of the Abbott Formation The Reynoldsburg Coal, near the base of the Abbott, is late Morrowan (latest Westphalian A) (fig. 6), on the basis of palynological study. The Tunnel Hill Coal, slightly above the Reynoldsburg and palynologically distinct from the Reynoldsburg, is classified as latest Morrowan (early Westphalian B stage). Conodonts indicative of late Morrowan or early Atokan age were recovered from limestone above the Tunnel Hill Coal in the Creal Springs Quadrangle (Jacobson 1983). The Oldtown Coal, which lies just above the Abbott in the basal Spoon Formation, is palynologically correlative with the Rock Island Coal and is classified as very late Atokan (latest Westphalian C). The Seville Limestone, which directly overlies the Rock Island Coal, contains fusulinids that also indicate late Atokan age. The Seville is considered equivalent to the Curlew Limestone of western Kentucky (Thompson et al. 1959, Douglass 1987).

Depositional environment Our preliminary findings, based largely on study of trace fossils, indicate that most shale and siltstone and much of the sandstone in the Abbott Formation were deposited under marine conditions. Marine body fossils have been found in the Creal Springs and Eddyville Quadrangles within the olive shale and equivalent sandstones of the middle part of the Abbott. Abundant, diverse ichnofossils are found in the same rocks in all three quadrangles. The forms *Asterosoma, Conostichus, Rhizocorallium, Teichichnus Torrowangia,* and *Zoophycus,* in particular, have strong marine affinities (Chamberlain 1971). The broad diversity of ichnogenera observed at many middle Abbott sites is an additional indicator of marine sedimentation (Miller 1984).

Environmental indicators are sparse in the basal Abbott shale and sandstone interval, but an occurrence of *Conostichus* in this interval and the presence of acritarchs in Reynoldsburg (?) coal indicate marine

32

deposition in these strata also. Within this framework, the thick-bedded, mappable Abbott sandstones seem best interpreted as distributary-channel or distributary-mouth bar deposits. The largest body of lower Abbott sandstone, which follows the McCormick Anticline from the eastern part of the Eddyville Quadrangle to the southwestern part of the Stonefort Quadrangle, probably represents a distributary channel. This sand body trends west to west-southwest; some of its features, including its erosional base, abundant trough crossbedding, and slumped laminations, are typical of distributary channels (Brown et al. 1973). Unfortunately, too few paleocurrent indicators were recorded to establish any trend. The location of this sandstone along the anticline suggests that structural movements may have controlled its deposition. Potter (1957) interpreted deposits immediately below this sandstone in the Illinois Central railroad cut as subaqueous gravity slides triggered by contemporaneous uplift of the McCormick Anticline.

The middle Abbott section thickens substantially and becomes sandier eastward in the study area. The

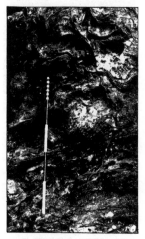

Figure 17 Liesegang banding, a common feature of thick-bedded to massive sandstone (especially in middle Abbott Formation), in Murray Bluff Sandstone at same locality shown in figure 16.

thickening may reflect accelerated tectonic subsidence of the area at the junction of the Reelfoot Rift and the Rough Creek Graben; this junction was a center of subsidence during much of Paleozoic time (Kolata and Nelson, in press). Deltas may have prograded into this area from the east, depositing mostly sand on the east and mud and silt offshore to the west. Some middle Abbott distributary channels, however, advanced as far as the Creal Springs Quadrangle. The best example of one of these channels is the south-trending linear body of Cedar Creek sandstone, with its unidirectional, south-facing foreset beds, in Section 20, T11S, R4E.

The thickest Murray Bluff Sandstone lies in a west- to southwest-trending belt parallel to the McCormick Anticline. Enclosed in shaly marine strata, it thins and becomes shaly as it approaches the flank of the anticline. Our own limited paleocurrent data, and data of Potter (1963, p. 46), indicate westerly to southwesterly flow. (Potter's map shows crossbed orientation of the "Finnie Sandstone" in the west-central part of the Stonefort Quadrangle. The outcrop pattern indicates that the sandstone is Murray Bluff.) The Murray Bluff fines upward from an erosional base and in places has slumped into underlying shale. All these features support interpretation of thick Murray Bluff as a distributary channel in a high-constructive, elongate delta (Brown et al. 1973). Contemporaneous structural movement on the McCormick Anticline may have influenced the course of the channel.

Spoon Formation

Name and definition Kosanke et al. (1960) proposed and named the formation for the Spoon River in Fulton County, west-central Illinois, where the type section was described. As originally defined, the Spoon comprises strata from the top of the Bernadotte/Murray Bluff Sandstone Member to the base of the Colchester Coal Member. We have revised the upper boundary of the Spoon Formation in the study area downward to the base of the Davis Coal Member for reasons explained in the section on name and definition of the Carbondale Formation.

Distribution and typography The Spoon Formation crops out along the northern edge of the Eddyville

Quadrangle and caps uplands in the Battle Ford Syncline west of Blackman Hollow. It occurs north and south of the New Burnside Anticline in the Stonefort Quadrangle and overlaps the crest of the anticline from Old Town nearly to the western edge of the quadrangle. The Spoon covers most of the northern three-quarters of the Creal Springs Quadrangle except the valleys of larger streams and an area along the New Burnside Anticline northeast of Sugar Creek, where it has been eroded. The Spoon is covered by glacial drift in the northwestern part of the Creal Springs Quadrangle and by Quaternary alluvial and lacustrine sediments in much of the northern part of the Stonefort Quadrangle.

A gently to moderately rolling topography developed in the Spoon Formation. Sandstones in the Spoon rarely produce ledges. Locally the sandstones cap low dissected plateaus, of which Wise Ridge and the hills north of Stonefort village are the best examples. Natural exposures of the Spoon are limited in size and quantity. The lithologic sequence is known mainly from subsurface data and exposures in quarries and strip mines.

Thickness The upper portion of the Spoon Formation is eroded in all but a small area of the Eddyville Quadrangle, so the original thickness of the formation within the study area is unknown. Boreholes north of the study area and composite measurements from outcrops indicate that the original thickness increased eastward from about 200 feet in the Creal Springs Quadrangle to 225 to 300 feet in parts of the Stonefort and Eddyville Quadrangles.

Lithology The Spoon Formation consists primarily of gray sandstone, siltstone, and shale (fig. 6) and generally contains a lower proportion of sandstone than the Abbott Formation in the eastern part of the study area. In the western part of the area the Abbott becomes shalier, so the contrast between the Abbott and Spoon is less marked.

Sandstones of the Spoon Formation, like most Abbott Sandstones, are classified as subgraywackes (Potter 1963, Potter and Glass 1958). Coarse mica flakes and carbonaceous plant fragment grains are conspicuous in most Spoon sandstones and thickly coat the bedding planes. The matrix is dominantly clay, and cement is weak or absent (Potter and Glass 1958). Thus, Spoon sandstones are less resistant and more friable than most older Pennsylvanian sandstones. Unweathered sandstone is light to medium gray or buff and commonly has a speckled or salt-and-pepper appearance due to carbonaceous grains. Weathered surfaces generally are yellowish gray to deep orange brown. The grain size varies from very fine to coarse. Small quartz granules are rare and found only in the lowest part of the Spoon. Shale, coal, and other sedimentary clasts are fairly common.

Shale and siltstone of the Spoon, as in the Abbott and Caseyville, vary from light gray to black. Clay-shale is more abundant in the Spoon than in older formations. Medium to dark gray clay-shales containing sideritic nodules are common.

Several coal beds in the Spoon are 3 to 4 feet thick and can be traced through large parts of the report area. Some of these coal beds may extend into several counties in southern Illinois.

Thin but widely traceable limestone and black fissile shale containing marine fossils are present in the Spoon Formation, particularly the upper part.

The following members, beds, and informal subdivisions of the Spoon Formation are recognized in this study (fig. 6):

Sub-Davis sandstone (youngest)

Strata between sub-Davis and golden sandstones
Carrier Mills Shale Member
Stonefort Limestone Member
Wise Ridge Coal Bed
Mt. Rorah Coal Member
Creal Springs Limestone Member
Murphysboro (?) Coal Member

Golden sandstone

Strata below golden sandstone
Mitchellsville Limestone Bed
New Burnside Coal Bed
Delwood Coal Bed
Oldtown Coal Bed (oldest)

Strata below golden sandstone
An interval of dominantly fine-grained strata, 30 to more than 130 feet thick, separates the Murray Bluff and golden sandstone. The Oldtown, Delwood, and New Burnside Coal Beds and the Mitchellsville Lime-

stone Bed occur within this basal interval of the Spoon Formation.

The basal Spoon Formation commonly is thin-bedded sandstone, gradational from the underlying Murray Bluff Sandstone. Ripple-laminated sandstone at the base commonly grades upward to gray siltstone or silty shale. The upper part of the Murray Bluff appears to grade laterally to thin-bedded sandstone and shale southeastward along Bill Hill Hollow and also southwestward from the type locality to the railroad cuts south of Oldtown. Thus, the Spoon and Abbott Formations interfinger laterally in this area through an interval that may be as thick as 100 feet.

The best exposures of basal Spoon strata are in the railroad cuts in E 1/2, Section 6, and NE, Section 7, T11S, R5E, Stonefort Quadrangle (fig. 10). At the base is 50 to 60 feet of thin-bedded, shaly sandstone, grading upward from the Murray Bluff Sandstone. This shaly sandstone is overlain by 40 to 45 feet of interlaminated, dark gray, silty shale and siltstone (fig. 18), having some thin sandstone interbeds, overlain in

Figure 18 Upward transition from thin-bedded, shaly sandstone to shale containing marine trace fossils, in basal Spoon Formation. This railroad cut is south of Oldtown, Stonefort Quadrangle in SW NW NW, Section 5, T11S, R5E. The staff is 5 feet (1.5 m) long. The trace fossil *Conostichus broadheadi*, a marine form, is common here.

34

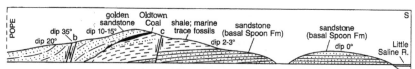

Figure 19 Profile of east side of Illinois Central railroad cut south of Oldtown, showing New Burnside Anticline, here a faulted monocline. (a) Fracture zone striking N80°W and dipping 90°, having horizontal slickensides and no measureable displacement. (b) Hinge of flexure; maximum dip about 35°N; two high-angle normal faults, each striking N75° - 80°W and having roughly 5 feet (1.5 m) of throw down to north. (c) Normal fault trending N60°E/60°NW, with about 10 feet (3.0 m) of throw. Not shown: many small, high-angle normal and some reverse faults or flexures, all having less than 1.0 feet (0.3 m) of throw; most are located between B and C or a short distance south of C. Location: along border of Sections 5 and 6, T11S, R5E.

overlying strata consist of medium to dark gray silty shale and siltstone and thin planar-bedded and ripple-laminated sandstone, along with lenticular intraformational conglomerates. The Oldtown Coal is 10 to 20 feet beneath the golden sandstone. The basal Spoon Formation is substantially thinner in the northeastern part of the Stonefort Quadrangle and in the Eddyville Quadrangle, where the Murray Bluff is thickest. The basal Spoon is as thin as 30 feet along the west side of Bill Hill Hollow, near the border of the two quadrangles. The interval is 40 to 60 feet thick along Battle Ford Creek and its tributaries farther east. As elsewhere, the interval generally becomes finer-grained upward and is in gradational contact to the Murray Bluff. A thin coal, overlain by shale containing plant fossils, was observed 10 to 20 feet below the base of the golden sandstone on the west side of Bill Hill, Hollow. We did not see any marine fossils between the Delwood Coal and the Murray Bluff Sandstone in the Eddyville Quadrangle. However, drillers' logs of coal test holes north of Delwood report limestone in this interval, and unpublished ISGS field notes record marine fossils, which we failed to locate during a specific search in Buzzard Roost Hollow.

Oldtown Coal Bed A coal bed up to about 3 feet thick crops out and has been mined at several places in the northern part of the Stonefort Quadrangle. We have named this coal the Oldtown Coal Bed. The type locality is the railroad cut just south of Oldtown (Old Stonefort) in SW NW NW, Section 5, T11S, R5E (figs. 10, 19). Kosanke et al. (1960) referred to the coal in the railroad cut as the Delwood Coal Member, which they placed in the Abbott Formation. Our mapping and palynological study demonstrate that the Oldtown Coal

is slightly older than the Delwood and closely correlates with the Rock Island Coal of northwestern Illinois, the Litchfield and Assumption Coals of central Illinois, and the Minshall Coal of Indiana.

The Oldtown Coal, as thick as 36 inches in the railroad cut (figs. 10, 19), is partly cut out by the overlying golden sandstone. M. E. Hopkins and Neely Bostick (ISGS, unpublished field notes, 1970) reported that just west of the railroad, in the E. and L. Coal Company surface mine, the coal was 34 to 35 inches thick and overlain by 4 to 10 feet of medium to dark gray, smooth, well-laminated shale, containing *Lingula* and plant fragments near the base. The golden sandstone overlies the shale.

The Oldtown Coal also was mined prior to 1916 at Frank Durfee's drift mine in SW SW SE, Section 2, T11S, R5E, western edge of Eddyville Quadrangle. At the time of mapping, the coal was partly exposed in the caved adits of this mine. A. D. Brokaw (ISGS, unpublished field notes, 1916) stated that the coal was 26 to 28 inches thick and directly overlain by sandstone (the golden sandstone). Smith (1957) reported 42 inches (?) of coal. West of Oldtown, coal float and adits or prospect pits occur in several places along the north flank of the New Burnside Anticline. The westernmost exposure of the Oldtown Coal is in the bank of Pond Creek, near the southwest corner of Section 3, T11S, R4E, Creal Springs Quadrangle. At this locality the Oldtown Coal is 8 to 12 inches thick and very shaly. Thin coal at widely scattered localities farther west possibly is equivalent to the Oldtown, but correlations could not be confirmed. Only underclay and a very thin coal occur at the expected position of the Oldtown Coal in boreholes E-1 and H-1 (plate 1), and in four proprietary test holes northwest

of Durfee's drift mine. The Oldtown Coal may be entirely eroded in these boreholes and the coal penetrated may be slightly below the position of the Oldtown. Table 5 lists results of chemical analysis from the Oldtown Coal at the E. and L. Mine.

Delwood Coal Bed Weller (1940) named the Delwood Coal for the village of Delwood; Wanless (1956) described the type section in NW NW, Section 3, T11S, R6E, Eddyville Quadrangle. These authors, as well as Kosanke et al. (1960), misidentified the golden sandstone as the Murray Bluff and assigned the Delwood Coal to the Abbott Formation. Our mapping, borehole data, and palynological study (by Peppers) demonstrate that the Delwood Coal is above the Murray Bluff Sandstone and that it is equivalent to the Bidwell Coal named by Kosanke et al. (1960). We believe the correlation of the two coals is strong enough to warrant abandoning the name Bidwell and retaining Delwood, which has priority. The Delwood is assigned bed status because it appears to be a local coal, not traceable very far beyond the present study area.

In most of the study area the Delwood Coal is 24 to 42 inches thick. The maximum thickness, 71 inches, is in a borehole in NE NE, Section 8, T11S, R4E, Creal Springs Quadrangle. This coal has been mined out. Forty-seven inches of coal was observed in a stream cut in NW NW NE, Section 4, T11S, R6E (Eddyville field station 110). The Delwood Coal crops out and has been mined at several places in the northeastern part of the Eddyville Quadrangle and in the Creal Springs Quadrangle northwest of New Burnside. Several boreholes, including hole H-1 (plate 1), have penetrated the coal in the southern Harrisburg Quadrangle (just north of the Eddyville Quadran-

35

Table 5 Analytical data on coal in the study area

Coal seam	Location	Fixed carbon (%)	Volatile matter (%)	Moisture (%)	Ash (%)	Sulfur (%)	Heating value, (Btu/lb)
		as-received basis					
Reynoldsburg	Mine dumps SW, 32-11S-4E	55.3	38.6	3.2	2.9	0.88	14,067
Reynoldsburg	Mine dumps NW SE SE, 31-11S-4E	51.8	40.3	4.2	3.7	1.90	14,045
Reynoldsburg	Herod Mine SE, 32-11S-4E	56.1	35.9	3.82	4.17	0.85	13,849
Reynoldsburg	Herod Mine SE, 32-11S-4E	53.4	39.6	3.44	3.53	0.66	14,077
Reynoldsburg	Herod Mine SE, 32-11S-4E	53.2	33.6	8.06	5.19	0.55	12,158
Reynoldsburg	Herod Mine SE, 32-11S-4E	54.1	38.9	3.20	3.82	0.95	14,109
Reynoldsburg	Herod Mine SE, 32-11S-4E	53.9	39.3	3.18	3.76	1.18	14,165
Reynoldsburg[a]	Ozark Mine SW NE SE, 31-11S-4E	52.0	37.6	3.4	7.0	3.22	13,525
Reynoldsburg (upper split)	Test hole S-4	52.3	33.7	1.0	13.0	10.08	12,354
Reynoldsburg (lower split)	Test hole S-4	38.7	28.5	1.0	31.8	6.22	10,375
Bell (?)[b]	Test pit SW SE SW, 31-11S-4E	45.3	32.8	16.7	5.2	0.60	10,488
Bell (?)[b]	Test pit SW SE SW, 31-11S-4E	44.7	33.3	17.6	4.4	0.63	10,344
Oldtown[c]	E. & L. Coal Co. Strip Mine, NE NE, 6-11S-5E	51.6	31.9	9.1	7.5	1.12	12,076
Oldtown[d]	Test hole S-3	47.3	28.6	1.9	22.2	1.86	11,152
New Burnside[c]	J. C. Jennings Mine SW NE NE, 8-11S-4E	48.6	37.8	8.6	5.0	2.12	12,971
Wise Ridge[e]	Test hole S-1 SW SE NE, 25-10S-4E	45.7	35.2	2.2	16.9	4.69	11,915
Mt. Rorah	Western Mining Co. SE SE, 34-10S-4E	46.7	32.7	3.6	17.0	7.73	—
Mt. Rorah[f]	Test hole S-1 SW SE NE, 25-10S-4E	41.3	29.6	2.2	26.9	8.27	10,025
Delwood[g]	Test hole S-2 SW NE SE, 29-10S-5E	53.6	30.6	5.1	10.8	2.65	12,241

[a] Composite of three face-channel samples.
[b] Weathered sample.
[c] Composite of two face-channel samples.
[d] Coal 0.4 foot thick.
[e] Coal 0.6 foot thick.
[f] Two claystone partings included in sample.
[g] One claystone parting included in sample.

gle). Test hole S-1 was cored in 42 inches of Delwood Coal in the north-central part of the Stonefort Quadrangle (plate 1). The Delwood Coal may have been eroded and replaced by the golden sandstone in the northwestern part of the Eddyville Quadrangle and in the outcrop belt in the Stonefort Quadrangle. A claystone parting occurs near the middle of the Delwood Coal at most sites. It is commonly 3 to 4 inches thick in the Eddyville Quadrangle,

and 6 to 18 inches thick in the Creal Springs Quadrangle. The parting is 30 inches thick in borehole H-1 (plate 1).

In the Eddyville and Harrisburg Quadrangles the Delwood Coal is overlain by 30 to 35 feet of dark gray shale, which contains plant fossils in the lower portions and coarsens upward to siltstone. This siltstone is overlain by the Mitchellsville Limestone. In the Creal Springs Quadrangle the Delwood Coal is overlain by 8 to 25 feet of shale, siltstone, and

sandstone overlain by the New Burnside Coal. The shale above the Delwood Coal is thin or absent in places because of erosion at the base of the golden sandstone. Table 5 gives analytical data on the Delwood Coal.

New Burnside Coal Bed The New Burnside Coal was named by Weller (1940). Its type section is west of New Burnside village, for which it was named, in SE SE SW, Section 5, T11S, R4E, Creal Springs Quadrangle. This

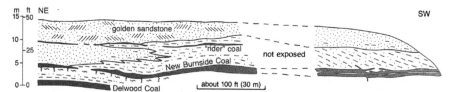

Figure 20 Drawing of surface-mine highwall (since reclaimed) in SW SE, Section 5, T11S, R4E, Creal Springs Quadrangle. New Burnside Coal is discontinuous and locally split with shale. The sequence above the New Burnside Coal, with sandstone on northwest grading laterally to siltstone and then silty shale with inclined foreset bedding, suggests a delta that prograded toward the southwest. The "rider" coal may be equivalent to the Murphysboro Coal Member, and is laterally equivalent to part of the golden sandstone. (From field sketches by John T. Popp 1977)

coal is restricted to the northeastern part of the Creal Springs Quadrangle. It has been mapped around the low hills west of New Burnside and in small areas of Section 4, T11S, R4E, and Section 32, T10S, R4E.

The New Burnside Coal is discontinuous and varies in thickness from a few inches to about 54 inches. Partings of shale, claystone, and bone coal are common. On strip-mine highwalls, the New Burnside Coal undulates, pinches, and swells. The strata overlying the coal also are highly variable (fig. 20). William H. Smith (ISGS, unpublished field notes, 1956) described sandstone with lenses of shale-pebble conglomerate up to 6 feet thick directly overlying the New Burnside Coal near the type locality. Kosanke et al. (1960, p. 65) noted conglomerates overlying both Bidwell (Delwood) and New Burnside Coals in their description of the type sections. Table 5 lists analytical data on the New Burnside Coal.

Mitchellsville Limestone Bed A distinctive, cherty marine limestone occurs about 35 feet above the Delwood Coal in the northern part of the Eddyville, and the southern part of the Harrisburg and Rudement Quadrangles (fig. 6). Although Butts (1925) and Weller (1940) called this rock the Curlew Limestone, Thompson et al. (1959) established, on the basis of fusulinids, that the

limestone of southern Illinois is younger than the type Curlew Limestone from Indian Hill, Union County, western Kentucky. Palynological study confirmed the conclusions based on fusulinids (Peppers and Popp 1979); therefore, we propose the name Mitchellsville Limestone Bed for the limestone formerly called "Curlew" in Illinois. The name is taken from the village of Mitchellsville, about 6 miles south of Harrisburg, Saline County. The type section of the Mitchellsville is in a north-trending ravine in SW NW NW, Section 27, T10S, R6E, Harrisburg Quadrangle (just north of the Eddyville Quadrangle). This ravine exposes about 5 feet of the limestone directly overlain by the golden sandstone. The lower contact of the limestone is concealed at the type locality; complete exposures have not been found. Cores from drillholes H-1 and E-1 (table 1), in permanent storage at the ISGS samples library, provide useful reference sections. The limestone is 1 foot thick at a depth of 41 to 42 feet in hole E-1, and 1.7 feet thick at a depth of 164.0 to 165.7 feet in hole H-1 (plate 1).

Fresh samples of Mitchellsville Limestone are medium to light gray or brownish gray. The limestone is sublithographic to fine grained and contains numerous crinoid fragments, corals, whole brachiopods, and other fossils. Bedding is nodular.

The limestone contains abundant large nodules of chert. It weathers to a yellow-brown porous residiuum containing hollow casts of brachiopods and other fossils. The cherty residiuum of the Mitchellsville is easy to recognize in float and thus the limestone can be mapped where outcrops do not exist.

In most places the Mitchellsville is directly overlain by the golden sandstone and has a contact that appears to be erosional. The limestone has not been found west of the Eddyville Quadrangle. Directly below the Mitchellsville at some sites is a very thin coal streak or coaly shale, with underclay beneath. The claystone grades downward to siltstone or shale, previously described in the section on the Delwood Coal.

Golden sandstone The golden sandstone* was called the Curlew sandstone in early reports, and renamed the Granger Sandstone by Kosanke et al. (1960). The type Granger Sandstone overlies the Curlew Limestone in Union County, Kentucky. Wanless (1939) correlated the "Curlew sandstone" of Kentucky with the "Granger sandstone" in southern Illinois. Correlation of sandstones was based on the assumption that the type Curlew Limestone was equivalent to Illinois "Curlew" (now called Mitchellsville). This correlation, is incorrect. The

*Comparison of maps and measured sections in the Creal Springs, Eddyville, and Stonefort Quadrangles indicates that the "golden sandstone" of the lower Spoon Formation represents at least two different sandstone bodies. What is described as the golden sandstone in this publication overlies the Delwood Coal along the north-central edge of the Eddyville Quadrangle; but in the Stonefort Quadrangle and in the northwestern part of the Eddyville Quadrangle, the sandstone mapped as "golden" probably underlies the Delwood Coal (not mapped)—a conclusion based on thickness considerations. In the Creal Springs Quadrangle, a sandstone occurs below the Delwood Coal and another sandstone occurs, at least locally, above the Delwood Coal; their positions are shown on the stratigraphic column, but their outcrops were not mapped in the quadrangle. The lower sandstone is equivalent to the golden sandstone as mapped in the Stonefort Quadrangle and the northwestern part of the Eddyville Quadrangle; the upper sandstone occurs in the same stratigraphic position as the golden sandstone along the north-central edge of the Eddyville Quadrangle. It has not been confirmed that the sandstones mapped as "golden" merge, as suggested by the stratigraphic columns of the Eddyville and Stonefort Quadrangles and in figures 6 and 20 of this report. Certainly, the lithology of all sandstone mapped as "golden" is similar.

informal name, golden sandstone, used in this report, refers to the golden brown color of the rock on some weathered surfaces.

The golden sandstone occurs in the syncline northeast of Delwood and can be seen in a roadcut on Route 145 near the center of the east line, E corner, Section 33, T10S, R6E, Eddyville Quadrangle. It also caps the drainage divide north and west of Rocky Branch and forms a cuesta along the northern edge of the quadrangle near Route 145. In the Stonefort Quadrangle the golden sandstone caps the hill between Caney Branch and Bill Hill Hollow and forms small hogbacks on the northern flank of the New Burnside Anticline. This is the sandstone that Kosanke et al. (1960) erroneously called the Murray Bluff in the Abbott type section (fig. 10). The golden sandstone was not mapped in the Creal Springs Quadrangle, but it crops out in many places. It is found near the tops of many hills north and west of New Burnside, and along the bed of Sugar Creek east of the village of Creal Springs.

Unweathered golden sandstone is light gray to buff; it weathers golden or orange brown to very dark brown, with coatings of iron oxide. Grain size ranges from fine to coarse with rare granules, and sorting is poor. This is the youngest sandstone in which quartz granules were observed. Mica is abundant; clay matrix, feldspar grains, and carbonaceous flakes are conspicuous. The sandstone tends to be friable and erodes to smooth, rounded forms. It seldom forms ledges, except in a few places near Murray Bluff, where the rock is pervasively impregnated with iron oxide. Bedding is generally thick to massive, with crossbedding common. In the northern part of the Stonefort Quadrangle the golden sandstone grades laterally to thin- or medium-bedded shaly sandstone, siltstone, and silty shale. Drill hole S-1, north of Stonefort village, encountered mainly siltstone at the expected position of the golden sandstone (plate 1).

The golden sandstone is about 20 to 50 feet thick in most places, and the top of the sandstone is eroded or covered nearly everywhere. It is at least 50 feet thick in the railroad cut. Kosanke et al. (1960) estimated 60 to 70 feet of golden ("Granger") sandstone near Creal Springs.

The contact of the golden sandstone with underlying strata is erosional. In the Eddyville and southern Harrisburg Quadrangles the golden sandstone overlies or fills an erosional channel cut into the Mitchellsville Limestone. Westward the sandstone cuts into the Oldtown Coal at Frank Durfee's drift mine (westernmost edge of the Eddyville Quadrangle) and along the outcrop belt in the Stonefort Quadrangle. The Oldtown Coal is partly eroded in the railroad cut south of Oldtown. Northward in the Stonefort Quadrangle, well data suggest that the golden sandstone may pinch out. In the Creal Springs Quadrangle the golden sandstone overlies the New Burnside Coal and cuts into it locally. The upper contact of the golden sandstone is gradational in the core of borehole H-1 (plate 1).

Strata between golden sandstone and sub-Davis sandstones The interval between the golden and sub-Davis sandstones consists largely of shale, siltstone, and claystone, with some shaly sandstone. Three named coal seams, two limestone members, and one newly named shale member occur within this interval. The total thickness of the interval varies from approximately 70 to 120 feet.

Murphysboro (?) Coal Member Coal probably equivalent to the Murphysboro Coal Member (Worthen 1868) formerly was mined in several small drift and strip mines west of New Burnside, in E 1/2 Section 8, T11S, R4E, Creal Springs Quadrangle. Identification of the coal is on the basis of its stratigraphic position and a palynological study of coal fragments collected from waste piles of mines. No outcrops of the coal were found in the study area. The coal occurs 20 to 30 feet above the New Burnside Coal and directly overlies the golden sandstone. On the highwall of a reclaimed strip mine west of New Burnside, we observed a thin "rider coal," possibly the Murphysboro, in a shale sequence that laterally interfingers with golden sandstone above the New Burnside Coal Bed (fig. 20).

Creal Springs Limestone Member The type locality of the Creal Springs Limestone Member is in an abandoned quarry about 1 mile east of Creal Springs village, in NE SE SE, Section 25, T10S, R3E (Kosanke et al. 1960). The limestone also is exposed in gullies north of Route 166 at the east edge of Creal Springs village; at the junction of an unnamed north-flowing stream and Brushy Creek in SE NE SE, Section 30, T10S, R4E; and in the bed of the north-trending ravine near the center N 1/2 SW, Section 28, same township. The limestone is less than 2 feet thick in most places. In the core of borehole C-4 it is 2.8 ft thick (plate 1). Lithologically the Creal Springs is similar to the Mitchellsville Limestone. The Creal Springs Limestone is medium to dark gray, argillaceous, and composed of whole brachiopods, gastropods, fusulinids, and echinoderm fragments in micritic matrix. Chert nodules are abundant and the limestone is totally silicified in places. The conspicuous chert float aids in mapping. The Creal Springs Limestone occurs above the golden sandstone. Its position relative to the Murphysboro (?) Coal west of New Burnside could not be determined.

In Franklin and Perry Counties, northwest of the study area, marine black shale and limestone overlie the Murphysboro Coal in drill cores. This limestone may be the Creal Springs, although continuity to the type area has not been established (Jacobson 1983).

Mt. Rorah Coal Member Kosanke et al. (1960) introduced the name Mt. Rorah Coal Member for a coal previously called the Bald Hill Coal. The name Mt. Rorah was taken from a church, now called the Mt. Moriah Church, near the northeastern corner of the Creal Springs Quadrangle. Kosanke et al. listed the type locality of the Mt. Rorah Coal as being in the SE, Section 35, T10S, R4E, but this appears to be a misprint. The coal does not crop out in this locality now, and no mention of it appears in ISGS field notes for the locality. The intended type locality probably is the ravine just east of the road at the northern edge of Stonefort village, SW NE SE, Section 25, T10S, R4E; the coal is no longer well exposed here either, but it is well described in ISGS field notes (open files).

The Mt. Rorah Coal is widespread in southern Illinois. It has been correlated in subsurface with coal in Perry and Jackson Counties (Jacobson 1983) and, less confidently, with

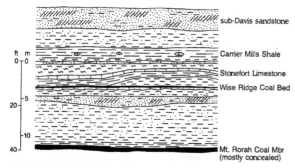

ft   m
0 — 0

          sub-Davis sandstone

          Carrier Mills Shale

          Stonefort Limestone

          Wise Ridge Coal Bed

—5

20 —

—10

40 —

          Mt. Rorah Coal Mbr
          (mostly concealed)

Figure 21 Strata exposed on highwall of reclaimed Western Mining Company surface mine, NE SE, Section 34, T10S, R4E, Stonefort Quadrangle. The Mt. Rorah Coal was mined here. Note lenticular body of light gray shale above the Wise Ridge Coal.

coal as far northeast as Lawrence County (Jacobsen 1987). In this report we have retained the Mt. Rorah Coal as a member rather than a bed because of its apparent continuity. The Mt. Rorah crops out and has been mined on a small scale in many places between the towns of Creal Springs and Stonefort. It is visible in the quarry at the type locality of the Creal Springs Limestone and in ravines northeast of Brushy Creek (Section 29, T10S, R4E), where it was dug in numerous small surface and drift mines. The coal also crops out around Wise Ridge and around the hills north and west of Stonefort. The Mt. Rorah Coal was mined in the reclaimed Western Mining Company strip mine, NE SE, Section 34, T10S, R4E (fig. 21). In addition, Mt. Rorah Coal was recovered in cores of drill holes H-1, S-1, and C-4 (plate 1).

Within the study area, the Mt. Rorah Coal ranges from less than a foot to about 45 inches thick, but it is commonly split with shale or claystone partings. Good exposures of the coal split by claystone are visible along the west-flowing stream just east of the center of Section 29, T10S, R4E, Creal Springs Quadrangle. The claystone partings contain coal stringers and range up to at least 12 inches thick. The coal in these exposures also is penetrated by several nearly vertical claystone dikes.

Little information is available on mines from which the Mt. Rorah Coal was extracted. The Western Mining Company pit was active during 1975 and covered an area of roughly 20

acres. No production figures are available. Nelson visited the site in 1977, after mining had ceased, and described strata in the highwall (fig. 21). The coal was largely concealed and the pit has since been backfilled. Three small surface mines are known in the Creal Springs Quadrangle. The Oxford Construction Company mined a small amount of Mt. Rorah coal in NE NE SW, Section 29, T10S, R4E during 1977. Two other small surface mines, dates and operators unknown, are in NW SW SW, Section 28 and NW SW NW, Section 33, both in T10S, R4E.

Cores and limited surface exposures indicate that the Mt. Rorah Coal is separated from the Creal Springs Limestone by 10 to 20 feet of medium gray, soft, smooth to silty shale, grading upward to rooted claystone. The coal generally is overlain by dark gray to black, silt-free, carbonaceous shale, which grades upward to gray silty shale, siltstone, and sandstone. Topping the sequence is the underclay of the Wise Ridge Coal. The Mt. Rorah to Wise Ridge interval is 20 to 26 feet thick. In the core of hole C-4 a dark gray to black, highly carbonaceous shale occurs about 6 feet below the Wise Ridge Coal Bed (plate 1).

Wise Ridge Coal Bed The Wise Ridge Coal, formerly called the Stonefort Coal, was named a member of the Spoon Formation by Kosanke et al. (1960) for Wise Ridge, a hill in the northeastern part of the Creal Springs Quadrangle. Its type locality is the same ravine that

contains the type Mt. Rorah Coal. The Wise Ridge Coal is thin, but fairly persistent across the northern edge of the study area. Its extent outside the mapped area is unknown; the coal is difficult to recognize in subsurface except in cores. It is therefore lowered in rank from member to bed in this report.

As far as we know, the Wise Ridge Coal does not exceed a thickness of 12 inches, and has not been mined. The few known exposures include several gullies on the hill northwest of Stonefort, and the quarry east of Creal Springs in NE SE SE, Section 25, T10S, R3E. The coal also occurs on the east side of Brushy Creek and on the north side of the hill east of Mt. Moriah Church. The Wise Ridge was exposed, prior to reclamation, on the highwall of the Western Mining Company surface mine, NE SE, Section 34, T10S, R4E (fig. 21). Wise Ridge Coal also was recovered in cores of drill holes C-4, H-1, and S-1 (plate 1).

In these cores and outcrops the Wise Ridge Coal is 6 to 12 inches thick, bright-banded, and blocky to thinly laminated. A soft, rooted claystone generally underlies the coal. A thin, shaly, nodular-bedded limestone was observed directly below the coal on the highwall of the Western Mining Company surface mine. Overlying the coal is 3 to 10 feet of moderately soft to firm dark gray, black or green-black, mottled, thinly laminated clay-shale, which is overlain by the Stonefort Limestone Member. On the highwall of this strip mine two distinct shales were observed above the Wise Ridge Coal. The lower unit was a soft, medium gray clay-shale in lenses up to about 2 feet thick directly above the coal. Sharply overlying the lower unit was about 3 feet of dark gray, thinly laminated shale (fig. 21).

Stonefort Limestone Member
Henbest (1928) named the Stonefort Limestone Member; Kosanke et al. (1960) established its type section in the same ravine (just north of Stonefort village) that contains the type Mt. Rorah and Wise Ridge Coals. The Stonefort Limestone, thin but widespread throughout southern Illinois, is readily identifiable on geophysical logs of oil test holes (Jacobson 1983; 1987). It is probably the most widely traceable and useful subsurface marker unit below the Davis Coal in

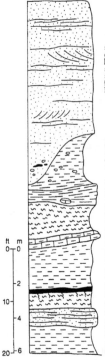

sub-Davis sandstone, light gray, weathers yellowish gray to orange-brown, very fine to fine, slightly friable, highly micaceous, slightly carbonaceous, much interstitial clay. Mostly thick-bedded, locally crossbedded; some thin-bedded portions with parallel to undulating bedding planes. Coal and shale rip-up clasts near base. Lower contact erosional.

shale, medium gray, silt-free to finely silty, well laminated, moderately firm, sideritic bands and nodules common. Becomes darker and softer near base, grades rapidly to underlying unit.

Carrier Mills Shale Member, black, smooth, hard, fissile, well jointed. Large septarian limestone concretions near base in places. Lower contact sharp and planar.

claystone, medium gray to olive gray, soft, not rooted. Zone of small, dark gray limestone nodules about 2 ft (0.6 m) from top. Lower contact sharp.

Stonefort Limestone Member, medium gray, very fine to fine, large shell fragments; shaly near top and base, upper contact undulates, lower contact planar.

shale, dark gray, silt-free, well laminated, lower part appears burrowed. Sharp contact.

shale, light to medium gray, soft, silt-free, moderately laminated, no fossils noted. Sharp contact.

Wise Ridge Coal Bed, bright to dull-banded, shaly, thinly laminated. Sharp contact.

claystone, medium gray to olive gray, soft, heavily rooted near top, lower part silty. Contact gradational.

sandstone, very light gray, very fine, friable, micaceous, argillaceous, ripply or lenticular thin to medium bedding. Sharp contact.

shale, light to medium gray, soft, weakly laminated, silt-free. Poorly exposed, base covered.

Figure 22 Section in abandoned railroad cut, NW NW NE, Section 30, T10S, R5E, Carrier Mills Quadrangle. This is the type locality of the Carrier Mills Shale Member of the Spoon Formation.

southern Illinois. In our study area the Stonefort is found in many of the same cores and surface exposures as the Wise Ridge Coal (figs. 22, 23; plate 1).

As observed, the Stonefort Lime stone varies from a few inches to a little less than 2 feet thick. It is medium to dark brownish gray, shaly, and fine grained with scattered coarse fossil fragments and whole fossils. The fauna includes brachiopods, pelecypods, gastropods, rugose corals, and crinoid fragments. The limestone typically forms a single bed or two beds separated by a shale parting. In some places it is nodular. The Stonefort Limestone does not leave a cherty residuum, as do the older Mitchellsville and Creal Springs Limestones.

About 5 to 15 feet of soft shale or claystone generally occurs between the Stonefort Limestone and the overlying Carrier Mills Shale. Shales and claystones above the Stonefort can be light gray to greenish gray, some mottled with red. They are commonly calcareous, and nodules of argillaceous limestone may be present. A thin bed of very shaly, nodular limestone was seen about 6 feet above the Stonefort on the highwall of the Western Mining Company surface mine (fig. 21), and in the core of borehole C-4 (plate 1). The nodular limestone and claystone, in core and highwall, showed evidence of rooting.

Carrier Mills Shale Member A distinctive hard, black, fissile shale, 2.5 to 5 feet thick, occurs 5 to 15 feet above the Stonefort Limestone and at or near the base of the sub-Davis sandstone, which has been mapped. This shale, which crops out in all

three quadrangles mapped, and in adjacent quadrangles north of the study area, also is widely traceable in subsurface. Jacobson (1987) has traced the black shale on geophysical logs through much of southern and east-central Illinois. This shale is therefore useful in both local and regional correlation. Jacobson (1987) noted that the shale appears to underlie the Seahorne Limestone Member of the Spoon Formation in west-central Illinois, but the correlation has not been confirmed.

We propose the name Carrier Mills Shale Member for the black shale. The name is taken from the town of Carrier Mills, located about midway between Stonefort and Harrisburg in southwestern Saline County (fig. 1). The type locality is an abandoned railroad cut in NW NW NE, Section 30, T10S, R5E, Carrier Mills Quadrangle. The Carrier Mills Shale in this locality is 2.5 to 4.0 feet thick and overlies soft, light gray claystone with a sharp contact (figs. 22, 23). The black shale grades upward to 1 to 10 feet of medium dark gray, silty, sideritic shale, which in turn is disconformably overlain by the sub-Davis sandstone. Reference sections of the Carrier Mills Shale (plate 1) are contained in cores of borehole S-1 (depth 69.5 to 73.8 ft), hole H-1 (depth 48.0 to 51.3 ft), and hole C-4 (depth 32.2 to 36.1 ft).

The shale is very dark gray to black, hard, and very thinly laminated. Unlike any older shale in the study area, it splits readily into sheets that are brittle to slightly flexible. Pyrite is common, and pyritized fossil fragments can be found. Fossils include *Lingula*, *Orbiculoidea*, and *Dunbarella*. The shale weathers to a silvery sheen, or occasionally to orange where pyrite is present. Because it is more resistant than adjacent gray shale, it generally projects slightly from banks or gully walls. Float of the Carrier Mills Shale is distinctive in many places just below the sub-Davis sandstone.

Thinly laminated, very shaly coal, or thin streaks of coal, occur locally at the base of the Carrier Mills Shale. The Carrier Mills Shale is overlain by as much as 10 feet of medium to dark gray, smooth to silty, well-laminated shale. In many places this shale is missing, and the sub-Davis sandstone lies directly on Carrier Mills Shale.

Figure 23 Carrier Mills Shale Member (behind staff) at the type locality (same locality shown in fig. 22). A thin layer of soft, gray shale overlies the hard, black Carrier Mills Shale; the sub-Davis sandstone disconformably overlies the gray shale. The Stonefort Limestone Member forms a broken ledge about 6 feet (1.8 m) beneath the Carrier Mills Shale and is separated from the latter by claystone. Staff is graduated in feet.

Sub-Davis sandstone The youngest recognizable unit of the Spoon Formation in our study area is a sandstone that caps many hills along the northern edge of the study area. The base of this sandstone is a mappable horizon, easily identified by the presence of the Carrier Mills Shale just beneath (figs. 22, 23, 24). We are calling it informally the sub-Davis sandstone, because it underlies the Davis Coal. The sub-Davis sandstone commonly is 30 to 40 feet thick, and locally it reaches 60 feet. The sub-Davis sandstone caps Wise Ridge and several other hills and divides in the northeastern Creal Springs Quadrangle. It also is believed to occur west of Creal Springs village, where it is concealed by glacial drift. All the hills northwest of U.S. Rt. 45 in the Stonefort Quadrangle are capped by sub-Davis sandstone. In the Eddyville Quadrangle the sandstone was mapped along the ravine southeast of Brown Brothers No. 1 Mine in NE, Section 30, T10S, R6E. The base of the sub-Davis sandstone is plotted on the Eddyville and Stonefort geologic maps.

Lithologically the sub-Davis sandstone resembles the golden sandstone, except that it lacks quartz granules. Large mica flakes and finely divided carbonaceous matter are concentrated on the bedding planes. The sandstone contains much feldspar and clay; it is friable

and weakly cemented. Grain size is fine to medium and locally coarse; sorting is poor. Lenses of conglomerate, composed of shale, coal, and siderite pebbles in a matrix of sandstone are common, especially at the base. Such conglomerates can be seen in the cores of boreholes S-1 and H-1 (plate 1). Bedding typically is thick; crossbedding is common.

The lower contact of the sub-Davis sandstone is sharp and erosional everywhere it has been observed. The upper contact is gradational to the underclay of the Davis Coal.

Age of the Spoon Formation The Oldtown Coal, near the base of the Spoon Formation, is of latest Atokan or latest Westphalian C age, on the basis of palynology (Peppers 1988). Douglass (1987) placed the Mitchellsville Limestone in the early Desmoinesian fusulinid zone of *Beedeina leei*. The Delwood Coal below the Mitchellsville Limestone contains spores and pollen, indicating Desmoinesian age. Apparently the boundaries of both the Atokan-Desmoinesian-Series and the Westphalian C and D Stages fall between the Oldtown and Delwood Coals in our study area (fig. 6).

Douglass (1987) placed the Stonefort Limestone Member (as sampled at its type locality) in the fusulinid zone of *Wedekindellina*, or early-middle Desmoinesian. Coals in the up-

per part of the Spoon Formation lie near the middle of the Desmoinesian Series and near the middle of the Westphalian D Stage on the basis of palynology (Peppers 1988).

Depositional environment Deposition of the Spoon Formation began with a pronounced transgression, which drowned the Murray Bluff delta. The transgression progressed eastward or northeastward, as shown by shaly strata that contain marine ichnofossils and body fossils, and interfinger with the upper part of the Murray Bluff Sandstone.

Following this incursion, a period of regression allowed widespread coal formation. Repeated minor fluctuations in base level allowed accumulation and burial of the Oldtown, Delwood, and New Burnside Coals, along with several thin, unnamed coals and rooted horizons. The patchy extent of coals suggests that the minor fluctuations were related mainly to local causes, such as structural movements or differential compaction of lenticular sediments at depth. For example, a basin of deposition developed in the Eddyville Quadrangle after accumulation of the Delwood Coal. This basin was filled successively by upward-coarsening clastics, a coal horizon or paleosol, and marine limestone (Mitchellsville). A much thinner clastic interval, and the New Burnside coal, were deposited at the same time in the Creal Springs Quadrangle, where no evidence of marine incursion exists.

The erosive base and rafted coal stringers of the golden sandstone suggest that it is a fluvial or distributary-channel deposit.

The widely persistent, uniform layering of the upper Spoon Formation reflects increasing regional stability, transitional to the widespread cyclic sedimentation of the Carbondale Formation. Two extensive marine incursions (Stonefort Limestone and Carrier Mills Shale) and a more localized incursion (Creal Springs Limestone) are indicated in the upper Spoon.

Carbondale Formation

Definition The Carbondale Formation was named for Carbondale, Illinois, by Lines (1912) and Shaw and Savage (1912). As originally defined, the Carbondale Formation comprised strata from the Murphysboro Coal,

41

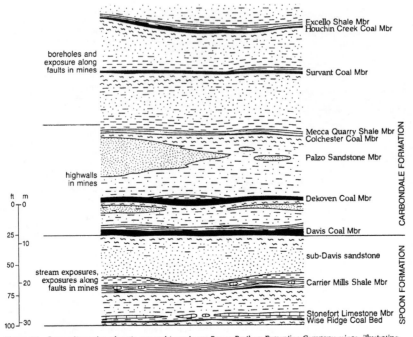

Excello Shale Mbr
Houchin Creek Coal Mbr

boreholes and
exposure along
faults in mines

Survant Coal Mbr

Mecca Quarry Shale Mbr
Colchester Coal Mbr

Palzo Sandstone Mbr

highwalls
in mines

Dekoven Coal Mbr

ft   m
0 — 0

Davis Coal Mbr

CARBONDALE FORMATION

25 —

sub-Davis sandstone

-10

50 —

stream exposures,
exposures along
faults in mines

-20

Carrier Mills Shale Mbr

75 —

SPOON FORMATION

Stonefort Limestone Mbr
Wise Ridge Coal Bed

100 ⌐ 30

Figure 24 Composite section of strata exposed in and near Brown Brothers Excavating Company mines, illustrating lateral variability (Sections 19, 20, and 30, T10S, R6E, Eddyville Quadrangle).

at the base, to the Herrin Coal at the top. Later, geologists who extended the Carbondale Formation away from its type area selected different boundary beds. The Carbondale Formation is currently recognized throughout the Illinois Basin, but its definition varies slightly from place to place (fig. 7).

Neither Lines nor Shaw and Savage designated a type section for the Carbondale, and suitable exposures probably do not exist in southern Illinois. Kosanke et al. (1960) designated a series of exposures near Canton, Illinois (more than 200 miles north of Carbondale), as a principal reference.

Within the current study area the Murphysboro Coal, originally defined as the basal member of the Carbondale Formation, is discontinuous and of doubtful identity. The oldest widely continuous and minable coal in the area is the Davis Coal,

which has been traced throughout southern Illinois and western Kentucky and correlates with the lower bench of the Seelyville Coal in west-central Indiana (Jacobson 1987). Moreover, the Davis Coal in southeastern Illinois marks the base of a succession characterized by widely continuous coal beds, each typically overlain by coarsening-upward clastics that commence with black fissile marine shale and terminate with sandstone and the underclay of the next coal. The regularity of this succession contrasts with the local variability of coals and inter-coal strata below the Davis Coal.

Therefore, we have adopted the base of the Davis Coal as the base of the Carbondale Formation in our study. This necessitates moving the Spoon-Carbondale Formation boundary proposed by Kosanke et al. (1960) about 70 to 80 feet downward from the base of the Colchester Coal

Member (fig. 6). The Colchester is the oldest widely mappable and minable coal bed in the area of west-central Illinois where Kosanke et al. defined the Spoon Formation. It remains to be seen whether the definition of the Spoon Formation should change in western Illinois, or whether a different classification should be introduced there.

The upper boundary of the Carbondale Formation lies at least 8 miles north of the present study area and is not an issue in this study.

Extent The Carbondale Formatio has been mapped in a downwarped area at the northern edge of the Eddyville Quadrangle in NE, Sectio 30, T10S, R6E. Lowermost Carbondale strata possibly occur elsewher atop hills rimmed by sub-Davis sandstone, but they do not crop out. Carbondale Formation occurs nort of the study area and is known there

42

in great detail from mine exposures and coal-test borehole data.

Thickness The maximum thickness of the Carbondale Formation within the Eddyville Quadrangle is approximately 225 feet, as estimated from boreholes. Another 175 to 250 feet of upper Carbondale strata has been eroded.

Lithology The Carbondale Formation has been removed by surface mining in the Brown Brothers No. 1 Mine southeast of the northeast-trending fault that runs through the center of Section 30, T10S, R6E. Geologic notes made while the mine was active, and on the recently closed Brown Brothers No. 2 Mine immediately to the north, indicate that the Davis and Dekoven Coal Members were mined. The Davis Coal measured 54 to 57 inches thick in the No. 1 Mine and 48 to 54 inches in the No. 2 Mine (fig. 24). The Davis was a bright-banded, blocky coal, with no significant shale partings. The floor was soft, rooted claystone, and the immediate roof was 2 to 3 feet of hard, black, fissile shale, similar to the Carrier Mills Shale. A coarse shell hash, composed of brachiopods and other marine fossils, commonly was found at the base of the shale.

Overlying the black shale was 20 to 25 feet of medium to dark gray, silty shale, medium gray siltstone, and light gray, shaly sandstone (fig. 24). The sequence was laterally and vertically variable. In some places it coarsened upward, but elsewhere sandstone was near the base. Capping the sequence was the underclay of the Dekoven Coal Member.

The Dekoven, 36 to 42 inches thick, was, like the Davis, clean, bright-banded, and free of notable partings.

The Dekoven Coal was overlain by approximately 50 feet of gray silty shale, siltstone, and sandstone. Dark gray to black silt-free, hard shale was found locally immediately above the Dekoven. Fossils in the shale included *Lingula, Orbiculoidea*, chonetid and productid brachiopods, and various pelecypods, including probable *Dunbarella*. Elsewhere the roof was medium to dark gray, smooth to silty sideritic shale or siltstone. In the upper half of the 50-foot interval was a sandstone up to about 20 feet thick (fig. 24).

The Colchester Coal was less than 12 inches thick in the Brown Brothers mines. Overlying the coal was 2 to 3 feet of black shale, grading up to gray shale or siltstone.

A small area of Carbondale Formation is mapped northwest of the No. 1 Mine in SW NE, Section 30, T10S, R6E, Eddyville Quadrangle. Identification of these rocks is based on test drilling by Brown Brothers and palynological analysis of coal exposed northwest of the fault zone on the highwall of the No. 1 Mine (fig. 24). A small area of coal believed to be the Houchin Creek Coal Member (formerly Summum Coal; Jacobson et al. 1985) was strip-mined immediately north of the Eddyville Quadrangle boundary by the Brown Brothers Company. The coal was about 24 to 30 inches thick, and overlain by shale, then sandstone. The Survant Coal, formerly Shawneetown Coal, (Jacobson et al. 1985), below the Houchin Creek, is known only from drilling records, which indicate a thickness of about 36 feet.

Age The portion of the Carbondale Formation within the study area is in the middle to middle-upper part of both the Desmoinesian Series and the Westphalian D stage (see fig. 6), on the basis of both fusulinids (Douglass 1987) and palynology (Peppers 1988).

43

SURFICIAL GEOLOGY

Steven P. Esling, Leon R. Follmer, Elizabeth D. Henderson,
Mary S. Lannon, and Matthew H. Riggs

The surficial deposits of the Creal Springs, Eddyville, and Stonefort 7.5-Minute Quadrangles were studied as a part of a cooperative project by the Department of Geology, Southern Illinois University at Carbondale (SIUC), and the Illinois State Geological Survey (ISGS). The surficial deposits include all the relatively soft, nonlithified sediments that cover the Paleozoic bedrock of the region. The deposits formed during the Quaternary Period, the most recent part of geologic time, including the present. The Quaternary deposits cover a major unconformity, a break in the geologic record that represents millions of years between the Quaternary and Paleozoic Periods.

During the Quaternary period, glaciers advanced from Canada to as far south as Creal Springs. In the study area the surficial deposits consist of loess, alluvium, colluvium, weathered residual material derived from bedrock, and minor amounts of glacial deposits (till and related sediments). Surficial materials can contain natural resources, including groundwater and sand and gravel, aggregates. Their composition, thickness, and physical characteristics determine the suitability of a site for specific land uses, such as a municipal landfill or industrial complex. The characteristics of surficial deposits also affect drainage, groundwater recharge, and the impact of certain environmental hazards, including flooding and earthquakes, on a region.

In this investigation we (1) characterized the surficial materials in terms of particle-size distribution, clay mineralogy, and thickness; (2) summarized the stratigraphy of the surficial deposits within the three quadrangles; (3) prepared detailed stack-unit maps of the surficial deposits in each of the quadrangles; (Riggs in prep., Henderson in prep., and Lannon in prep.); (4) assessed the resource potential of the Quaternary deposits, and (4) evaluated the suitability of the surficial materials for specific land uses.

PHYSIOGRAPHY

The Creal Springs, Eddyville, and Stonefort Quadrangles are in the Shawnee Hills Section of the Interior Lowland Province of Illinois (Leighton and others 1948). The area is maturely dissected and is dominated by narrow upland ridges. The highest point in the three quadrangles (approximately 840 ft above sea level) is just south of the town of Oak, along the eastern edge of the study area; the lowest point (about 380 ft above sea level) is where the Little Saline River and Pond Creek leave the study area at the north boundary. This gives a total relief of about 460 feet, and maximum local relief is about 240 feet. The study area contains the headwaters of the Saline River, Cache River, Bay Creek, and Lusk Creeks. All of these streams drain into the Ohio River; however, a prominent drainage divide trends east across the southern part of the quadrangles, separating streams flowing northward into the Saline River from those flowing southward into the Ohio River.

Most of the study area, except for the far northwest corner of the Creal Springs Quadrangle, lies south of the Illinoian glacial boundary (Willman and Frye 1980). However, north of the prominent drainage divide the valleys show the effects of glaciation. They generally are broader than valleys south of the drainage divide, and they are filled with sediment originating partly from processes associated with Illinoian and Wisconsinan glaciation. South of the divide the valleys are generally narrow, and the streams flow on rock or on thin alluvial deposits.

METHODOLOGY

Field Methods

Because exposed Quaternary sections are rare in southern Illinois, samples of the surficial materials were collected with a hydraulic soil probe (Giddings Machine Company). The soil probe recovers core and auger samples 6 to 15 meters (20 to 50 ft deep. Samples were generally collected with a slotted tube, yielding cores approximately 7.6 or 5 centimeters in diameter, or with a flight auger 5 centimeter in diameter. We selected 87 drilling sites on the basis of landscape position and accessibility and looked for stable positions where complete stratigraphic sequences might be preserved.

Cores were described in detail under field-moisture conditions. Munsell color, mottles, soil structure, texture, presence or absence of carbonate minerals, manganese oxide and iron oxide concretions, and other features characteristic of a particular stratigraphic unit were described. The texture of the deposits was determined by the U.S. Department of Agriculture textural classification. The weathering zone terminology used is that of Follmer, Tandarich, and Darmody (1985). Geologic samples were collected at 10- to 15-centimeter intervals from the cores and saved for laboratory analysis.

Laboratory Methods

The particle-size distribution of the samples was determined at the SIUC Department of Geology, using a pipette method described by Graham (1985). Particle-size classes include clay (less than 0.002 mm); fine silt (0.002 to 0.02 mm), coarse silt (0.02 to 0.62 mm), and sand (0.062 to 2 mm).

Clay mineralogy of the samples was determined at SIUC by X-ray analysis of orientated slides of the less than 0.002-mm fraction. The clay fraction was placed on glass slides, dried, glycolated, and analyzed for peak heights of standard minerals. The clay mineral suite recognizes peaks in the region of 17Å as expandable clay minerals, 10Å as illite, and 7Å as kaolinite and chlorite. Graham (1985) described a method, modified from Hallberg and others (1978b), which is based on a routine method used by the Illinois State Geological Survey to scan large numbers of samples for stratigraphic and classification purposes. This method yields

45

useful clay mineralogy data for characterizing Quaternary sediments; however, many factors other than clay mineralogy may affect peak height, and the accuracy of the method is not well documented. Therefore, conclusions based on comparisons between clay mineralogy data from this report and clay mineralogy data determined by similar methods on different instruments should be used with caution.

Construction of the Stack-Unit Map

The distribution of surficial deposits (lithostratigraphic units) is a function of the depositional conditions and the postdepositional erosional history of each unit. These two factors have caused complex relationships among the stratigraphic units in the study area. A stack-unit map (Kempton 1981) illustrates the areal and vertical distribution of geologic materials to a given depth below the surface. A generalized stack-unit map was prepared by combining information obtained from 87 borings made in the area, field observations, and hand augering. Principal sources of additional information were the county soil survey reports (Miles and Weiss 1978, Parks 1975, Fehrenbacher 1964, Fehrenbacher and Odell 1959) and well logs on file at the Illinois State Geological Survey. Soil reports provided information on the distribution and nature of the near-surface materials. Maps included in the county soil reports show the distribution of particular soil series in the study area and are useful in distinguishing major landscape and geologic boundaries. The surficial lithostratigraphic units were interpreted from the soil map units (table 6).

The preparation of a stack-unit map requires some subjective decisions: in many cases, map boundaries cannot be drawn with certainty because of insufficient data. The vertical sequence of deposits from a borehole is first related to the soil series and landscape position; topographic features are then used as a guide to drawing map-unit boundaries. A boundary between geologic units can cross a soil-map boundary for two reasons: (1) the soils maps and the 7.5-minute topographic base of the stack-unit maps are at different scales, and the topographic base does not show the subtle changes in slope identified on the soils maps;

46

Table 6 Relation between modern soil series and map units of this study

UPLAND SOILS

Unglaciated areas	Glaciated areas
Alford (generally slopes less than 12%)	Ava (generally slopes less than 12%)
Grantsburg (generally slopes less than 12%)	Bluford (slopes less than 4%)
Hosmer (generally slopes less than 12%)	Grantsburg (generally slopes less
Robbs	than 12%)
Stoy	Robbs

SIDESLOPES

Unglaciated areas	Glaciated areas
Alford (generally slopes more than 12%)	Ava (generally slopes more than 12%)
Alford-Baxter Complex (generally slopes more than 12%)	Grantsburg (generally slopes more than 12%)
Beasley	Hickory
Bedford	Hickory-Ava Complex
Berks	Wellston
Creal	Zanesville
Grantsburg (generally slopes more than 12%)	
Hosmer (generally slopes more than 12%)	
Muskingum	
Wellston	
Wellston-Berks Complex	
Wellston-Muskin Complex	
Zanesville	

LOWLANDS

Banlic	Burnside
Belknap	Creal
Bonnie	Sharon

ROCK OUTCROPS

Sandstone Rock Land
Wellston-Berks complex

and (2) in geologic mapping, data from deep borings must be integrated with near-surface data. In general, geologic material more than 5 feet below ground surface has little effect on the type of soil series at a site.

A detailed stack unit map for each quadrangle was prepared at a scale of 1:24,000 (Henderson in prep, Lannon in prep, Riggs in prep). These maps are useful for general planning purposes, but do not provide sufficient information for site-specific interpretations. These three maps will be placed in the ISGS Open File Series when they are completed.

STRATIGRAPHY

Stratigraphic studies facilitate understanding of the origin of the units and provide a framework for correlating and predicting their character and occurrence. The relationships of the surficial deposits of the Creal Springs, Eddyville, and Stonefort Quadrangles can be classified as (1)

chronostratigraphic (time relationships), (2) lithostratigraphic (lithologic or compositional relationships), and (3) pedostratigraphic (soil development history or relationships of pedogenic attributes).

Most of the stratigraphic units in the area correlate with units described by Willman and Frye (1970) and Lineback (1979). One new informal lithostratigraphic unit, the Oak formation, is proposed; this formation is a widespread weathered material (principally residuum) that directly overlies the bedrock in the region. Because many of the characteristics of the Oak formation are pedogenic in origin, this formation presents a classification problem; however, it has lithologic properties on which to base correlations and occupies a consistent stratigraphic position throughout the region.

The distribution of lithostratigraphic units in the study area is controlled to a large extent by geomorphic processes of erosion and eolian deposition; this explains the

CHRONOSTRATIGRAPHY				LITHOSTRATIGRAPHY			PEDOSTRATIGRAPHY
QUATERNARY SYSTEM	Pleistocene Series	Wisconsinan Stage	Holocene Stage				Modern Soil
			Valderan/Greatlakean Substage	Peyton Fm	Cahokia Fm		
			Twocreekan Substage				
			Woodfordian Substage	Peoria Silt	Equality Fm (Mbr B)		
			Farmdalian Substage				Farmdale Soil
			Altonian Substage	Roxana Silt	Equality Fm (Mbr A)		
		Sangamonian Stage					Sangamon Soil
		Illinoian Stage		Glasford Fm	Teneriffe Silt	Loveland Silt	
		Pre-illinoian		Oak fm			Yarmouth Soil

Figure 25 Stratigraphic classification of Quaternary deposits in the study area. Chronostratigraphic names are commonly used for the deposits (substage) that formed during an interval of time (subage).

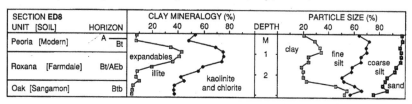

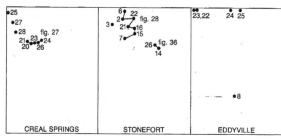

Figure 26 Vertical distribution of clay minerals and particle sizes in boring ED8, and general locations of borings mentioned in this chapter.

distinction between the sequence of surficial deposits constituting the present upland and the sequence in lowland positions.

Lithostratigraphic Units: Upland Formations

Oak formation This formation is dominated by fine-grained material with a soil fabric (peds or natural aggregates and biological pores); it overlies bedrock on stable upland and some hillslope surfaces and

underlies some valley deposits. The Oak formation is weathered material; most of it is the product of in situ weathering, but it includes redeposited, stratified material in places. All other Quaternary units overlie it (fig. 25). We propose that the informal name, Oak formation, be used to replace a variety of common names (i.e., residual material, weathered bedrock, paleosol, geest, clay, and pedisediment), and that the Oak formation be recognized as a minor unit

in lithostratigraphic classification. Because a residuum has attributes of both a soil and a lithologic (material) unit, it has often been excluded from lithostratigraphic classification. The Oak formation in the study area is laterally continuous and it has lithologic properties distinctly different from those of the underlying indurated bedrock. The term *residuum* is restricted in this discussion to material formed in situ on a bedrock substrate. The term *formation* may include in situ and redeposited material.

The Oak formation is named for the town of Oak in Pope County, near the eastern boundary of the Eddyville Quadrangle. A boring (ED8 in the Eddyville Quadrangle) that contains a typical example of the Oak formation serves as a reference section. In this section the Roxana Silt overlies the Oak formation, which rests on bedrock. Figure 26 summarizes laboratory data for ED8, and appendix A (page 59) describes the ED8 section. General locations of borings discussed in this chapter are shown in figure 26; precise locations are listed in table 7.

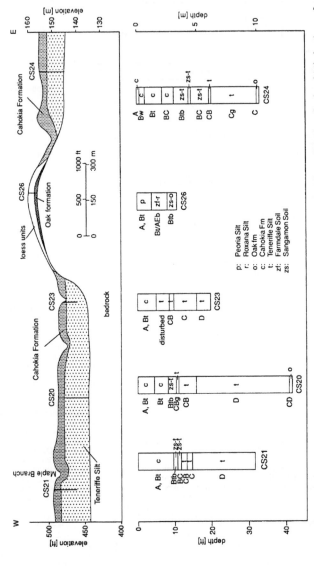

Figure 27 Cross section showing the distribution of Quaternary units in the Sugar Creek valley (Creal Springs 7.5-minute Quadrangle) (for location see fig. 26).

48

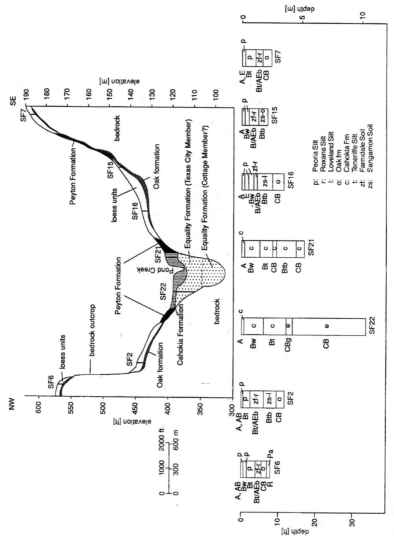

Figure 28 Cross section showing the distribution of Quaternary units in the Pond Creek valley (Stonefort 7.5-minute Quadrangle) (for location, see fig. 26).

49

Table 7 Location of borings

Boring	Location		Elev. (ft)
Creal Springs Quadrangle			
CS20	SW NE SE	2-11S-3E	485
CS21	NW NW SE	2-11S-3E	495
CS23	SE NE SE	2-11S-3E	485
CS24	NW NW SE	1-11S-3E	510
CS25	NW NW SW	27-10S-3E	590
CS26	SW NE SW	1-11S-3E	525
CS27	SW NE NW	34-10S-3E	615
CS28	NW SW SE	34-10S-3E	545
Eddyville Quadrangle			
ED8	SE NW NE	29-11S-6E	845
ED22	SE SE NE	26-10S-5E	395
ED23	SE SW NE	26-10S-5E	383
ED24	SW SE NW	29-10S-6E	390
ED25	NW SW NW	28-10S-6E	370
Stonefort Quadrangle			
SF2	SE SW SE	25-10S-4E	450
SF3	SW SE NE	6-10S 4E	490
SF6	SW SE NE	25-10S-4E	570
SF7	NE SE NE	1-11S-5E	620
SF1	NW NW NW	9-11S-5E	400
SF1	SE SE SW	31-11S-5E	480
SF1	SE NE SW	31-10S-5E	442
SF21	NE NW SW	31-10S-5E	405
SF22	SE SE SW	30-10S-5E	393
SF26	NW SE SE	5-11S-5E	405

The Oak formation is generally continuous on uplands of low relief. On valley slopes it is discontinuous and occasionally exposed; it may occur under alluvium in valleys. The Oak formation is uncemented and softer than the underlying bedrock. The lower boundary, gradational in most cases, is defined by a distinct change in consistence or an increase in hardness. The upper contact is an unconformity that ranges from abrupt to gradational into the overlying younger units that commonly shows contrast in color, fabric, and texture. On the uplands in the study area the Oak formation is overlain by the Loveland Silt, the Glasford Formation, or the Roxana Silt. In the valleys, it is overlain by the Cahokia Formation, the Equality Formation, or the Teneriffe Silt.

The Oak formation is 0.66 meters thick at the reference section (appendix A) and ranges from 0.3 to 2.7 meters thick across the study area.

The texture (particle-size distribution) reflects the composition of the underlying bedrock, ranging from loam and sandy loam to sandy clay loam where it overlies sandstone, and from silt loam and clay loam to clay where it overlies shale. The high concentrations of clay and sand show the effect of pedologic alteration and the predominance of sandstone lithologies of the bedrock in the study area. The sand and pebble content generally increase with depth. The unit is leached of carbonates in all but one observed location in a valley. In most cases the unit is oxidized, and reddish brown, brown, and yellowish brown hues are common. The Oak formation is characterized by a high percentage of kaolinite and chlorite clay minerals, reflecting the clay mineralogy of the parent rock. At some locations the percentage of expandable clay minerals is high near the top of the unit, indicating possible mixing with the overlying Roxana Silt. In other regions the residuum on carbonate rocks is rich in expandable (smectitic) clays, such as the geest in northwestern Illinois described by Willman et al. (1989).

The Sugar Creek and Pond Creek transects, summarized in figures 27 and 28, show the physical relationship between the Oak formation and the other valley and upland Quaternary-age units in the study area. The Oak formation is not shown in the cross sections of valleys because of scale limitations and lack of data. A type of material included in the Oak formation found in valley positions is the unoxidized and unleached sediment overlying bedrock in the Maple Branch of Sugar Creek Valley (fig. 29).

Figures 30, 31, and 32 summarize the vertical trends in particle size and clay mineralogy of borings SF3, CS25, and CS27. These borings serve as additional reference sections for the Oak formation.

Interpreting the age of a lithostratigraphic unit that accumulated primarily from weathering processes presents a unique classification problem. As the weathering front progresses and the contact between the lithified and nonlithified units moves downwards, the residuum becomes thicker. Therefore, the softened part of the bedrock becomes residuum. In addition, when the residuum is at ground surface, it can thicken upward as new deposits are

added and weathered. The residual material on bedrock probably develops in episodes; it could be eroded during periods of landscape instability and could form again during periods of stability. We have not determined when Oak development began. It could have started when the Pennsylvanian rocks were first exposed to weathering more than 200 million years ago, but we have no evidence that it is older than late Tertiary (several million years old).

Landscape evolution models suggest that in regions having appreciable relief, erosion would remove or rework all residual material in less than 2 million years. Residual materials older than 2 million years occur in a few areas in North America and are recognized as remnants of lateritic weathering. Lateritic iron ore or bauxite occurs at the top of deep weathering profiles where erosion has not occurred. Rare observations of lateritic material have been noted in Illinois, Indiana, Arkansas, and a few places in the western part of the United States and in Canada. No lateritic materials have been found in the study area.

The Oak formation is not weathered to the degree of a laterite. It contains a mixed clay-mineral assemblage that includes expandable clay. The landscape evolution models and the degree of weathering in the Oak suggest that most of the formation is of Quaternary age. Most of the Oak appears to be about 300,000 to 2 million years old, on the basis of its stratigraphic occurrence below correlated Illinoian units (Loveland Silt, Glasford Formation, and Teneriffe Silt). Although the Oak directly underlies the Wisconsinan deposits in many places, it often appears to be an erosional remnant of older, deeply weathered material. The residuum would thicken between times of erosion, and the actual time of formation is not a critical factor for mapping its distribution.

The Oak formation usually shows pronounced pedologic features, including well-developed soil fabric (structure) and clay films on ped faces (coatings on aggregates), the evidence for interpreting an old soil or paleosol. Where overlain by the Roxana Silt, this paleosol is correlated with the Sangamon Soil; where overlain by the Loveland Silt, the paleosol is correlated with the Yarmouth Soil. Although the paleosol in

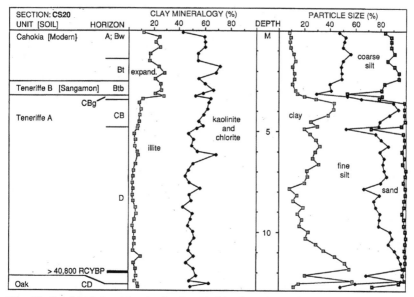

Figure 29 Vertical distribution of clay minerals and particle sizes in boring CS20 (for location see fig. 26).

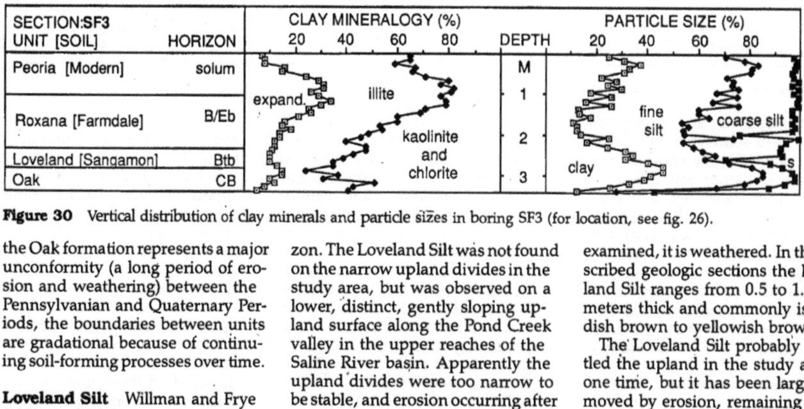

Figure 30 Vertical distribution of clay minerals and particle sizes in boring SF3 (for location, see fig. 26).

the Oak formation represents a major unconformity (a long period of erosion and weathering) between the Pennsylvanian and Quaternary Periods, the boundaries between units are gradational because of continuing soil-forming processes over time.

Loveland Silt Willman and Frye (1970) applied the term Loveland Silt to the undifferentiated silt deposits of Illinoian age that occur south of the Illinoian glacial limit. The Loveland Silt is primarily loess. It overlies the Oak formation and is buried by the Roxana Silt. Sangamon Soil development has altered the entire unit, creating blocky structure and biological pores typical of a B horizon. The Loveland Silt was not found on the narrow upland divides in the study area, but was observed on a lower, distinct, gently sloping upland surface along the Pond Creek valley in the upper reaches of the Saline River basin. Apparently the upland divides were too narrow to be stable, and erosion occurring after deposition of the Loveland Silt removed most of the deposit. Detailed logs of borings SF2 and SF16 (fig. 28) show the Loveland Silt on the landscape positions adjacent to Pond Creek. Although the lower upland position is broader than the divides, the Loveland Silt was encountered in only three of the eight borings drilled at this position. At all locations examined, it is weathered. In the described geologic sections the Loveland Silt ranges from 0.5 to 1.25 meters thick and commonly is reddish brown to yellowish brown.

The Loveland Silt probably mantled the upland in the study area at one time, but it has been largely removed by erosion, remaining as a deposit of colluvium and loess only on protected, stable landscape positions. Weathered Loveland Silt and Roxana Silt are similar in appearance and composition and and can be easily differentiated only when the Sangamon Soil is recognized in the Loveland Silt.

The texture of the weathered Loveland Silt ranges from a silt loam to

51

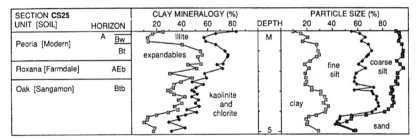

Figure 31 Vertical distribution of clay minerals and particle sizes in boring CS25 (for location, see fig. 26).

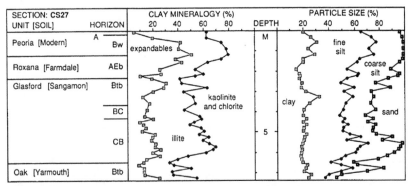

Figure 32 Vertical distribution of clay minerals and particle sizes in boring CS27 (for location, see fig. 26).

silty clay loam, and its clay content increases somewhat with depth within the Bt (clay-enriched) horizon of the Sangamon Soil. The sand content also increases with depth as a result of mixing with the underlying Oak formation. The Loveland Silt in the study area contains high percentages of kaolinite plus chlorite clay minerals. Figure 30 summarizes the vertical distribution of clay minerals and particle size of a typical Loveland Silt section (SF3) in the study area.

Glasford Formation The Glasford Formation mainly consists of Illinoian-age diamicton (a mixture of clay, silt, sand, and pebbles). Diamicton that is dense and has uniform characteristics is interpreted as till, a sediment directly deposited from a glacier. Other types of diamicton include deposits that have been locally redeposited by water action or mass wasting. The Glasford Formation is the most extensive Quaternary lithostratigraphic unit in Illinois, and its

southern boundary marks the southernmost extent of continental glaciation in North America (William and Frye, 1980). In the study area the Glasford Formation is found only in the extreme northeast corner of the Creal Springs Quadrangle. Where present, it overlies the Oak formation and is buried by the Roxana Silt. The development of the Sangamon Soil has altered all of the Glasford Formation in upland positions; glacial deposits may fill bedrock valleys, and unweathered diamicton might be found in these valleys.

The map compiled by Lineback (1979) shows the Vandalia Member, a middle member of the Glasford Formation, covering the study area. Willman and Frye (1970) interpreted the southern limit of the Glasford as forming during the oldest part of the Illinoian. The limited information acquired in this study has not resolved this correlation question. Therefore, we classify all observed glacial deposits in the study area as

undifferentiated Glasford Formation and reserve the correlation question for further investigation.

In the study area the diamicton appears to have a patchy distribution. Only two of the four borings north of the glacial boundary encountered the unit. The deposit is highly weathered and thin, and in both described sections the unit is oxidized to a brown or yellowish brown and leached of carbonates. In boring CS27 (fig. 32), the Glasford Formation is 3.4 meters thick. The base of the Glasford Formation was not encountered in boring CS28, from which 3.3 meters of diamicton was recovered.

The Glasford Formation in the study area generally has a silt-loam texture but ranges to silty clay loam and loam. Erratic igneous and metamorphic pebbles are rare, and the coarser clasts are mainly sandstone, coal, and chert derived from the local bedrock. Typically, the Glasford Formation contains high percentages

52

pedologic alteration from modern soil development. Expandable clay minerals dominate the clay fraction near the top of the Roxana and gradually diminish with depth as the kaolinite plus chlorite fraction increases. This vertical trend may indicate a change in source area as well as mixing of the Roxana Silt and the underlying Oak formation. Frye and others (1962) attributed the high percentage of expandable clay minerals to a northwest source area. Figures 26, 30, 31, and 32 summarize typical vertical trends in particle-size distribution and clay mineralogy of the Roxana Silt in the study area.

Peoria Silt The Peoria Silt, also called Peoria Loess, was deposited during the Woodfordian Subage of the Wisconsinan Stage by aeolian processes (McKay 1979); it is the surficial deposit on the upland of the study area, overlying the Roxana Silt in most places. The Peoria Silt is thinner in the study area (0.5 to 1.3 m) than in locations near major rivers, where it can be more than 6 meters thick. It is typically altered to yellowish brown and is lighter in color than the Roxana. Its physical appearance is dominated by soil characteristics (humic enrichment, soil structure, and biological features) and it is leached of carbonate minerals in all locations examined.

Like the Roxana Silt, the Peoria Silt is characterized by consistent vertical trends in clay mineralogy and particle-size distribution. These vertical trends show the effect of modern pedogenesis; most profiles show a distinct Bt (clay-enriched) horizon of a forest-type soil (Alfisol). The Peoria Silt contains more silt and less sand than does the underlying Roxana Silt; its texture ranges from silty clay loam to silt loam. Near ground surface, the Peoria contains more illite than expandable clay minerals. With depth the percentage of expandable clay minerals increases, and kaolinite plus chlorite generally decreases. This vertical trend is probably a function of three factors: pedogenesis, provenance change, and mixing with the Roxana Silt. The trends in clay mineralogy are typical for the southern Illinois region (Graham 1985, Hughes 1987). Figures 26, 30, 31, and 32 graphically summarize the vertical distribution of clay minerals and particle size of typical Peoria Silt profiles in the study area.

Peyton Formation The Peyton Formation includes all sediment deposited by mass wasting and colluvium at the base of slopes that has accumulated from the end of the Wisconsinan to the present. The Peyton consists of reworked loess, residuum, and clasts of bedrock. The unit, generally thin, is commonly associated with bedrock outcrops. The Peyton generally is yellowish brown to brown and is leached of carbonate minerals in the study area.

The Peyton colluvial material on slopes grades into alluvium that is stratified and better sorted in the valley bottoms. An arbitrary vertical line separates deposits that originated mostly from fluvial processes (the Cahokia Formation) from those derived mainly from colluvial processes (the Peyton Formation.) The Sugar Creek and Pond Creek transects (fig. 27, 28) show the physical relationship between the Peyton Formation, the loess units, and the valley deposits in the study area.

Lithostratigraphic Units: Valley Formations

Teneriffe Silt The Teneriffe Silt is a term applied in the study area to a laterally continuous, stratified deposit in the Sugar Creek valley in the northwest part of the Creal Springs Quadrangle, just south of the Illinoian Glacial boundary. The deposit is correlated with the Teneriffe Silt on the basis of the presence of Sangamon Soil in the upper part of the unit and its facies relationship with the adjacent Glasford Formation. The Illinoian-age Teneriffe Silt overlies either bedrock or Oak formation and is buried by the Cahokia Formation (fig. 27).

No natural exposures of the Teneriffe Silt exist in the study area. Sufficient evidence was found in four borings to subdivide the formation into two informal members. The lower part of the Teneriffe (member A) is a poorly stratified silt, typically characterized by a silty clay texture, although silt loam, silty clay loam, sandy clay loam, sandy loam, loam, and clay textures are also present. The upper part of the Teneriffe Silt (member B) is characterized by a high percentage of sand in silt loam, loam, sandy loam, or clay loam textures. Member B is brown to yellowish brown and is leached of carbonates. Member A may be yellowish brown and brown near its top, but becomes

53

bluish gray, greenish gray, or gray, and calcareous with depth. Pedologic development has altered all of member B and the upper part of member A. The resulting soil, interpreted to be the Sangamon Soil, formed in the Teneriffe and is more than 3.3 meters thick in one boring (CS24). It has a distinct Bt (clay-enriched) horizon, suggesting cumulic development over an extended period of time. The high percentage of fine particle sizes in member A (clay content up to 66% in some beds) suggests that the Teneriffe Silt accumulated in a low-energy, quiet-water environment. In the two borings that penetrated the Teneriffe, the unit is 6.9 and 10 meters thick.

The clay mineralogy of the Teneriffe in four borings reveals no consistent patterns that characterize the unit. In general, the Teneriffe Silt in the study area contains mainly illite and kaolinite plus chlorite. The assemblage of clay minerals can be explained by a mixing of materials from two principal source areas: adjacent diamicton from the Illinoian glacier, and older materials from the watershed of Sugar Creek. Borings near tributary streams draining the area where the Glasford Formation is present contain high percentages of illite, whereas those located near streams that drain the Oak formation have high percentages of kaolinite plus chlorite. Materials containing high percentages of kaolinite plus chlorite have a high sand content. The Oak formation also contains a high percentage of sand and kaolinite plus chlorite. In three of the four borings, the percentage of expandable clay minerals is greatest near the top of the Teneriffe Silt. Figures 29 and 33 summarize the vertical distribution of clay minerals and particle size in two Teneriffe Silt sections.

Three criteria suggest that the unit described in the Sugar Creek valley correlates with the Teneriffe Silt: the Sangamon-class paleosol in the upper part of the unit; the high percentage of illite (a characteristic of Illinoian-age glacial deposits in the region); and a radiocarbon date from a conifer sample found at a depth between 11.91 to 11.98 meters in boring CS20 (figs. 27, 29). This wood sample was dated at >40,800 yr BP (ISGS 1476). Three factors (the confinement of the deposit within the Sugar Creek valley in the study area, the proximity of the Glasford Formation, and

the high percentage of silt and clay) suggest that the Teneriffe Silt accumulated in an ice-marginal Illinoian lake. Willman and Frye (1980) discussed the origin of this ice-marginal lake, which they named Lake Sugar, and presented geomorphic evidence suggesting that the Illinoian glacier blocked the Sugar Creek Valley just to the east of the town of Creal Springs. The elevation at the top of the Teneriffe Silt in the Maple Branch of Sugar Creek valley is below 500 feet. Willman and Frye (1980) believed that the bedrock surface of an outlet to Lake Sugar at an elevation of about 520 feet controlled the maximum elevation of the lake.

A silt loam deposit correlated with the Cahokia Formation overlies the Teneriffe Silt. The surface of the Teneriffe Silt is an unconformity, spanning Sangamonian through early Wisconsinan time. The Roxana and Peoria Silts would have been deposited over the Teneriffe; however, more information is needed to distinguish these silt deposits. Subsequent Wisconsinan and Holocene fluvial erosion and deposition has made differentiation of the Cahokia Formation from the Roxana or Peoria Silts problematic in the four described sections containing the Teneriffe Silt.

Correlation of this Illinoian silt with the Teneriffe Silt is based on a precedent established by Lineback (1979). He applied the term Teneriffe Silt to silts, clayey silts, sands, and clays deposited in basins marginal to the Illinoian glaciers. The Teneriffe Silt, as mapped by Lineback (1979), is recognized beyond the margin and on top of the Glasford Formation. In its statewide use, the Teneriffe Silt may include eolian, glaciofluvial, glaciolacustrine, or mass wasted deposits. Heinrich (1982) applied the name Teneriffe Silt to Illinoian-age lacustrine deposits within the Saline River valley just to the northeast of the study area.

It is important to recognize that Lineback's definition of the Teneriffe Silt differs from the original definition. Willman and Frye (1970) excluded the Teneriffe Silt from areas beyond the Illinoian glacial boundary, preferring to draw a vertical cutoff between the Loveland and Teneriffe Silts. They interpreted the Teneriffe Silt to include outwash, sheetwash, and some eolian deposits, without mentioning ice-marginal

lacustrine deposits. Proper characterization of the important variations in these deposits will require further investigation.

Equality Formation Willman and Frye (1970) interpreted the Equality Formation as Woodfordian-age lacustrine deposits associated with Wisconsinan glaciation. The type section for this unit occurs in the Saline River valley, just to the northeast of the study area. The Equality Formation in the study area is described in two borings in the Battle Ford Creek valley, two borings in the Little Saline River valley, one boring in the Pond Creek valley, and one boring in the Grassy Creek valley. Shaw (1915), Frye and others (1972), and Heinrich (1982) developed a conceptual model for deposition of the Equality Formation in the Saline River drainage basin. During the Altonian and Woodfordian Subages of the Wisconsinan, the Ohio River carried glacial outwash and aggraded faster than its tributary valleys having drainage basins south of the glacial border. High water levels and sediment dams formed at the mouth of the Saline River, impounding water within the drainage basin. Sediment carried by floodwaters from the Ohio River or by streams originating on the adjacent uplands accumulated in the lake. In the study area the Equality Formation is mostly a silt loam, but it may have silty clay loam, sandy loam, or loam textures. The unit, almost always unoxidized, can be gray, dark gray, greenish gray, or dark greenish gray. The Equality Formation (except for beds near the top of the unit within the Battle Ford Creek valley) is generally leached of carbonates.

The distribution of clay minerals within the Equality Formation appears to be a function of location in the Saline River drainage basin. Heinrich (1982) found a high percentage of illite in the Equality Formation in the Saline River valley northeast of the study area. These deposits originated partly from the illite-rich sediment eroded from the Glasford Formation or carried in suspension into the Saline River valley by floods on the Ohio River. The deposits in the Battle Ford Creek valley, in the upper reaches of the Saline River drainage basin, contain high percentages of kaolinite plus chlorite, reflecting a local bedrock provenance.

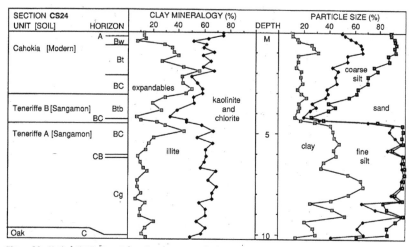

Figure 33 Vertical distribution of clay minerals and particle sizes in boring CS24 (for location, see fig. 26).

In the Grassy Creek, Pond Creek, and Little Saline River valleys, the Equality Formation contains a high percentage of expandable clay minerals, suggesting that the sediment was partly derived from the Roxana Silt that mantled the uplands during the early Woodfordian. Figures 34 and 35 summarize the vertical distribution of clay minerals and particle size of borings within the Little Saline River Valley. The upper part of the Equality Formation in this valley is characterized by a high percentage of sand.

Heinrich (1982) subdivided the Equality Formation into two informal members, the Altonian Subage Cottage member and the Woodfordian Subage Texas City member, which he differentiated by stratigraphic position and composition. He determined that during the Altonian the Cottage member accumulated in a lake (impounded in the Saline River valley), which reached a maximum elevation of about 350 feet. He suggested that the maximum elevation of the lake during the Woodfordian was about 380 feet. Deposits of the Equality Formation described from borings in the present study probably correlate with the Texas City member, because deposits of the Cottage member would not be found in tributary valleys of the Saline River above an elevation of 350 feet. Only one of the borings (ED25) penetrated

below 350 feet to a minimum elevation of 329 feet, but data from this core are largely inconclusive. The Pond Creek and Little Saline River transects (figs. 28, 36) show the distribution of both the Texas City and Cottage members, but the contact between the members is hypothetical (the contact is placed at Heinrich's (1982) approximate maximum elevation for the Altonian lacustrine episode. In boring SF26 the contact between the Cahokia and Equality Formations is at an elevation of about 387 feet, which suggests a lake level during the Woodfordian greater than Heinrich's (1982) estimate of 380 feet. The elevation of this contact supports the estimate by Frye and others (1972) of about 400 feet for the maximum level of the lake during the Woodfordian. The Maple Branch portion of the Sugar Creek valley in the Creal Springs Quadrangle does not contain deposits belonging to the Equality Formation. The contact between the bedrock and the valley fill in this valley is probably at an elevation above the maximum lake levels reached during the Wisconsinan. Because only a few borings in the quadrangles encountered the Equality Formation, no evidence was found for subunits that could correlate with the textural and mineralogic subunits described by Heinrich (1982).

Willman and Frye (1970) separated the Equality Formation into the Carmi Member and Dolton Member. By this classification the deposits in the study area correlate with the Carmi Member, which includes the predominantly fine-textured deposits that accumulated in low-energy environments. The coarser end of the spectrum is assigned to the Dolton Member, typically sandy, which represents higher-energy environments such as beaches.

In the study area, no evidence for a paleosol was found in the upper part of the Equality Formation beneath its contact with the overlying Cahokia Formation. A paleosol was described below this contact at other locations in southern Illinois (Graham 1985; Oliver 1988), suggesting that in the study area erosion has removed it. In boring SF14 (fig. 34) and SF26 (fig. 35) the contact between the Cahokia and Equality Formations is placed at a distinct change in texture. In boring SF14, the contact is placed at the base of a fining-upward sequence. In boring SF26, the contact is placed at the upper contact of a prominent sandy loam unit. In two of the borings in the Battle Ford Creek valley (ED24 and ED25), the boundary between the Cahokia and Equality Formations is problematic. In boring ED25, the contact is placed at a depth of 8.4 meters, where both

55

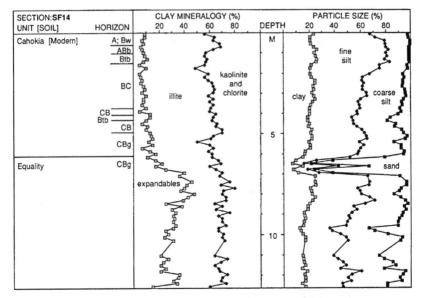

Figure 34 Vertical distribution of clay minerals and particle sizes in boring SF14 (for location, see fig. 26).

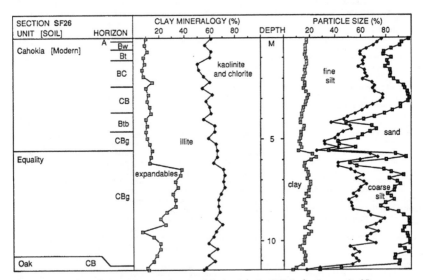

Figure 35 Vertical distribution of clay minerals and particle sizes in boring SF26 (for location, see fig. 26).

the clay content and illite content increase. Organic material retrieved at the base of ED24 is dated at >21,000 yr BP (ISGS 1456). Frye and others (1972) and Heinrich (1982) believed that water was impounded along the Saline River valley during the Woodfordian. Thus, some of the material above the base of boring ED24, which has a basal elevation of about 356 feet, may belong to the Equality Formation. A tentative boundary is placed at a depth of 5.4 meters. Below this depth the laminated, calcareous sediment contains more clay size material and illite.

Cahokia Formation Willman and Frye (1970) interpreted the Cahokia Formation as Holocene-age alluvial deposits. The unit is the surface deposit in all the valleys of the study area except the narrow tributary valleys in which streams flow on rock. Pedologic features, including soil fabric and evidence of translocated clay, extend throughout the entire thickness of this unit; these features are characteristic of a cumulic soil. The texture of the Cahokia indicates that it originated from the loess or redeposited loess. The Cahokia Formation may overlie the Oak formation, the Equality Formation, or the Teneriffe Silt. The thickness of the Cahokia Formation is highly variable, ranging from 0 to 8.4 meters. Most borings terminated before the base of the unit was penetrated. The variable thickness of the Cahokia Formation is evidence for cycles of aggradation and degradation in the tributary valleys of the study area during the Holocene. The Sugar Creek, Pond Creek, and Little Saline River transects (figs. 27, 28, 36) show the physical relationship between the Cahokia Formation and the other upland and valley stratigraphic units.

Near the surface the Cahokia Formation is generally oxidized to various shades of brown and yellowish brown that grade into unoxidized gray and bluish grays with depth. The unit is leached of carbonates in all described geologic sections. The Cahokia Formation is silty; its textures are similar to its source materials, the upland loess deposits and residuum. In general, the unit has a silt loam texture but may also consist of loamy sand, sandy loam, silty clay loam, silty clay, and loam.

Although its clay mineralogy varies from one location to the next, the Cahokia Formation generally contains a high percentage of kaolinite plus chlorite and illite, in roughly equal amounts. With few exceptions, expandable clay minerals constitute less than 20 percent of the total clay mineral content; this suggests that the Peoria loess provided most of the sediment, but the distribution of clay minerals was probably a function of the clay mineralogy of the exposed upland units within a local drainage basin. The Cahokia Formation does not have distinct vertical trends in either particle-size distribution or clay mineralogy. However, over short distances, beds within the unit may be correlated. For example, borings ED22 and ED23 are separated by about 4,000 meters and show similar trends with increasing depth. Figures 29, 33, 34, and 35 graphically summarize the vertical distribution of clay minerals and particle size for typical sections in the study area. Wood recovered at a depth of 5 meters from boring ED25 was dated at 5,500 ± 80 yr BP (ISGS 1457).

Pedostratigraphic Units

Yarmouth Soil In the study area, the Yarmouth Soil developed in the Oak formation and is overlain by the Glasford Formation or the Loveland Silt. The Yarmouth is distinguished from the other soil-stratigraphic units in the area by its stratigraphic position. In other regions the Yarmouth is thicker and more weathered than the Sangamon Soil, but in the study

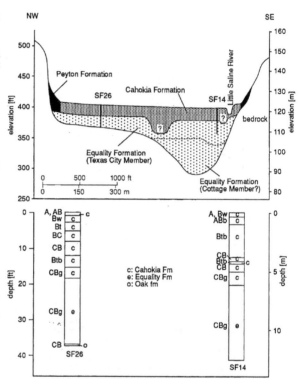

Figure 36 Cross section showing the distribution of Quaternary units in the Little Saline River valley (Stonefort 7.5-minute Quadrangle) (for location, see fig. 26).

57

area the differences are hard to distinguish.

Sangamon Soil The Sangamon Soil in the study area developed in parts of Glasford Formation, Loveland Silt, Teneriffe Silt, and Oak formation exposed at land surface during Sangamonian time. The Sangamon Soil is distinguished from other pedologic units by its stratigraphic position and the strength of its development; in many places it has a pronounced Bt horizon and moderate to strong blocky fabric that is commonly called soil structure.

The Sangamon Soil is buried by Wisconsinan-age Roxana Silt on the uplands in the study area. Except in the Sugar Creek Valley, the Sangamon Soil preserved in valleys is developed in the Oak formation and buried by the Wisconsinan-age Equality Formation; in the Sugar Creek Valley, the Bt horizon of the Sangamon Soil is developed in the Teneriffe Silt and is buried by the Holocene-age Cahokia Formation.

Farmdale Soil The Farmdale Soil in the study area formed in the Roxana Silt and was buried by the Peoria Silt. The Farmdale Soil is easily identified by its thin to medium platy structure, although modern pedogenesis has obscured some of its primary features. Silt coats or masses of clean, white silt grains commonly occur along the ped faces. Many biological pores appear to be primary features, but the many clay films, manganese and iron oxide concretions, and stains along root channels probably reflect overprinting of the Farmdale Soil by modern pedogenesis.

RESOURCES

Surficial deposits in the study area may contain a variety of natural resources such as groundwater, gravel, sand, and clay. Silt and clay-rich materials in some areas may be suitable for ceramic products.

No significant groundwater resources have been found in surficial deposits of the study area. The upland deposits are generally unsaturated, and the valleys contain fine-textured sediments that seldom yield groundwater in economic quantities. Isolated sand lenses within the valley deposits may yield sufficient groundwater to supply a single-family dwelling, but these units are generally located only by chance. The study area does not contain economic deposits of sand and gravel aggregate within the surficial materials. The weathered silt deposits that cover the upland have little commodity value except for construction fill or ceramic products such as common brick.

LAND USE

The value of the surficial deposits of a region in terms of material and space for various land uses is becoming increasingly important; it is in these near-surface materials that roads, houses, and industrial facilities are built, sewers and pipe lines are laid, crops are raised, and wastes are buried, and the value of land, particularly in developing regions, is directly related to its suitability for such uses.

In the three quadrangles, important land-use issues (beyond agricultural considerations) include the siting of waste disposal (landfill) operations and major construction projects. The characteristics and thickness of the surficial deposits are critical to determining the suitability of a site for a particular type of land use. Although no large municipalities are located in the study area at present, growth could occur in the region in future years. Once a general area is selected for any prospective construction or waste disposal use, a detailed on-site evaluation must be made of the surface deposits before the facility can be designed and constructed. This section summarizes the basic characteristics of surface materials in the quadrangles that can affect waste disposal, construction practices, and seismic risk considerations related to such projects.

Waste Disposal

Municipal waste includes household refuse and nonsalvageable commercial wastes such as metal and paper. This type of refuse is commonly buried in sanitary landfills. The primary consideration in locating a sanitary landfill is minimizing the migration of leachate (a solution produced when infiltrating water reacts with the refuse) into groundwater and surface water. Guidelines of the Illinois Department of Public Health (1966) require that in selecting a sanitary (municipal) landfill site (1) no waste can be disposed of in standing water; (2) no waste can be disposed of in areas with a high water table unless preventive measures are taken to prevent leachate migration; (3) no surface runoff should flow into or through the operation or completed fill area; and (4) no disposal should take place "unless the subsoil material affords reasonable assurance that leachate from the landfill will not contaminate groundwater or surface water."

In general, no site within the Creal Springs or Stonefort Quadrangles is suitable for a sanitary landfill unless measures are taken to prevent the migration of leachate. The upland loess units over the nonglaciated areas of these two quadrangles are less than 2.5 meters thick, which is too thin for disposal trenches or for use as a source for trench-cover material. In the glaciated area, the Glasford Formation has a patchy distribution, and the total loess and diamicton thickness is rarely more than 7 meters. No major bedrock aquifers are located in the study area, but the upland units generally overlie sandstone that serves as a minor aquifer in the Creal Springs and Stonefort Quadrangles.

Suitable waste disposal sites may exist in the Eddyville Quadrangle. In this quadrangle, shale, a bedrock unit having low hydraulic conductivity, underlies the surficial deposits on most upland positions. The Roxana Silt and Peoria Silt have a combined thickness of less than 2 meters in this quadrangle, however, providing only a marginal supply of cover material. Valley units of all three quadrangles are located in regions having poor surface drainage and high water tables (generally within 1 meter of the surface in most bottomlands); therefore, the valleys are not suitable for sanitary landfills. In places where no other site is available within a reasonable distance of a community, a sanitary landfill could be located in the study area if the landfill is lined and leachate is collected and treated.

Even more stringent geologic conditions are required for the safe disposal of hazardous wastes, including toxic chemical, biological, radioactive, flammable, and explosive refuse. No suitable sites for hazardous waste disposal were found in the study area.

General Construction

Most general construction projects, such as roads, subdivisions, small businesses, drainage systems, and water and sewage lines, can, if properly designed and built, be sited anywhere in the study area except in lowlands prone to flooding or on the extremely steep valley slopes. Conditions on the uplands are most favorable. Loess, a silty deposit, typically has a low bearing capacity and is susceptible to frost action if water saturated; however, the loess covering the upland areas in the study area is unsaturated for most of the year and probably has a moderate bearing capacity. Sandstone bedrock underlying the surficial material in many upland positions has a high bearing capacity and is generally found at shallow depths (less than 2.5 m). The shallow bedrock may be advantageous for projects requiring high bearing capacity, but it is a liability in excavating for sewer lines, water lines, or septic systems. The glacial diamicton that underlies the loess in the northwest part of the Creal Springs Quadrangle has a variable texture, but in most places, this material has properties comparable to those of the loess. Detailed study of lateral variation in the material properties of the diamicton is necessary before any construction project is begun.

The valley bottoms in the study area are generally not suitable for construction. A high water table (less than 1 m in places), poor surface drainage, and silty materials having low bearing capacity characterize these areas. Flash flooding is also a distinct possibility in the narrow valleys.

Seismic Risk

The study area lies in a zone of high seismicity. Historical evidence suggests that earthquakes of large magnitude, associated with the New Madrid Fault Zone in eastern Missouri, could occur. In the winter of 1811-1812, a series of major earthquakes shook the region, and intensities of IX or X (on the Modified Mercalli Intensity Scale) were probably experienced in the three quadrangles; earthquakes of this intensity could cause total destruction of weak structures and major damage to well-built structures (Nuttli 1973).

Structures built in the valley bottoms of the area are most at risk; the thick, unconsolidated, and saturated deposits found in these areas generally amplify seismic waves and are prone to liquefaction, which causes a significant loss of bearing capacity. Structures having foundations in bedrock are at less risk. The foundations of critical structures such as hospitals, schools, and emergency service centers should be located in the bedrock on gently sloping uplands.

APPENDIX A   Description of Boring ED8

Description of a core sample collected from a borehole on a narrow upland interfluve located 5 meters north of County Road 23 and 1.1 kilometer east of Illinois State Highway 145 (SE NW NE, Section 29, T11S, R6E, Eddyville 7.5-minute Quadrangle, Pope County). Approximate surface elevation is 845 feet.

Depth	Zone	Description
Peoria Silt with Modern Soil		
0.00-0.28	A, E	A and E horizon of Modern soil, brown (10YR 5/3) silt loam, granular, porous and friable; gradual boundary.
0.28-0.61	Bt1	Yellowish brown (10YR 5/6) silty clay loam; moderate, fine sub-angular blocky structure; few, fine manganese oxide stains along root channels; abrupt boundary.
0.61-0.99	Bt2	Yellowish brown (10YR 5/6) silty clay loam with common, medium light grayish brown (10YR 6/2) mottles and common, medium, light gray (10YR 7/1) silt coats and common, fine brown (10YR 5/3) clay films; moderate fine subangular blocky structure; common, fine to medium, dark yellowish brown (10YR 4/4) iron oxide concretions; large, very dark brown (10YR 2/2) root cast with gray (10YR 5/1 and 7.5YR 6/1) clay films; abrupt boundary.
Roxana Silt with Farmdale Soil		
0.99-1.75	2Bt/AEb	Yellowish brown (10YR 5/6) silt loam with common, medium dark brown (10YR 4/3) mottles and common thin to moderately thick light brownish gray (10YR 6/2) clay films and common light gray (10YR 7/1) silt coats; strong, thin platy structure; few, fine manganese oxide stains along root channels; diffuse boundary.
1.75-2.28	2Bwb1	Yellowish brown (10YR 5/6 and 5/4) silt loam; disturbed sample; common, fine to medium manganese oxide concretions; clear boundary.
2.28-2.39	2Bwb2	Yellowish brown (10YR 5/6 and 5/4) silt loam; strong, fine angular blocky structure; common, medium manganese oxide concretions; abrupt boundary.
Oak formation with Sangamon Soil		
2.39-3.05	3Btb	Yellowish brown (10YR 5/6) and yellowish red (5YR 5/6) clay with common, medium brown (10YR 5/3) to dark brown (10YR 4/3) clay films; strong fine angular blocky structure; common dark yellowish brown (10YR 4/4), dark reddish brown (5YR 3/3), and yellow (10YR 7/8) sandstone pebbles; gradual boundary; leached residuum.
3.05	3R	Bedrock; Pennsylvanian sandstone.

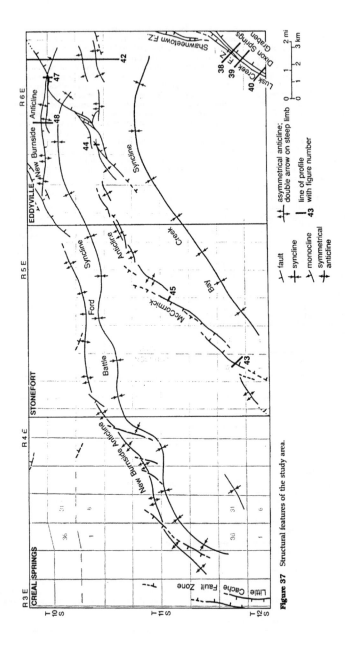

Figure 37 Structural features of the study area.

60

STRUCTURAL GEOLOGY AND MINERAL DEPOSITS

W. John Nelson

The three quadrangles of interest lie along the southern margin of the Illinois Basin (fig. 2). The strata dip regionally northward toward the structural center of the basin, which is located about 60 miles north of the study area. The average dip is roughly 1 in 50 (1°) in the Eddyville and Stonefort Quadrangles and slightly less in the Creal Springs Quadrangle. Regional dip direction is slightly west of north in the Eddyville Quadrangle, due north in the Stonefort Quadrangle, and slightly east of north in the Creal Springs Quadrangle.

The structural pattern is strongly modified by four important zones of deformation that cross the study area (fig. 37): the Shawneetown Fault Zone, which grazes the eastern edge of the Eddyville Quadrangle; the Lusk Creek Fault Zone, which crosses the southeastern corner of the same quadrangle; the McCormick Anticline, which curves westward from the east-central part of the Eddyville Quandrangle to the southwesternmost part of the Stonefort Quadrangle; and the New Burnside Anticline, which runs westward across the northern part of the Eddyville and Stonefort Quadrangles and turns southwest through the Creal Springs Quadrangle. The newly named Battle Ford Syncline is between the McCormick and New Burnside Anticlines, and the newly named Bay Creek Syncline is south of the McCormick Anticline. The Little Cache Fault Zone, also newly named, is in the southwestern corner of the study area.

STRUCTURAL FEATURES

Shawneetown Fault Zone

The Shawneetown Fault Zone is part of the Rough Creek-Shawneetown Fault System, which runs about 125 miles east-west across southeastern Illinois and western Kentucky and has maximum vertical displacement of at least 3,500 feet (Nelson and Lumm 1984). Near its western end, the Shawneetown Fault Zone curves sharply to the southwest and

intersects the Lusk Creek Fault Zone near the eastern edge of the Eddyville Quadrangle in Section 25, T11S, R6E (fig. 37).

The only part of the Shawneetown Fault Zone within the study area is a pair of faults forming a narrow graben that strikes slightly east of north in W1/2, Section 24, and NW, Section 25, T11S, R6E. Faulting is indicated by steeply dipping, shattered, and slickensided rock and by juxtaposition of pebbly Caseyville sandstone with brown, micaceous sandstone believed to belong to the Abbott Formation. The faults apparently converge and die out toward the south. Northward, they extend out of the quadrangle and merge with the main part of the Shawneetown Fault Zone, as mapped by Baxter et al. (1967) in the Herod Quadrangle. The faults apparently are high angle, but their direction of dip is unknown. The dips of rocks adjacent to the faults are not consistent from place to place. The strata on both sides of the faults dip west in the north part of Section 24, but farther south they dip eastward. The reversed dips suggest the possibility of more than one episode of movement.

Lusk Creek Fault Zone

The Lusk Creek Fault Zone was named for Lusk Creek by Weller et al. (1952). It trends southwest from the western terminus of the Shawneetown Fault Zone and crosses the southeastern corner of the Eddyville Quadrangle. From there, the fault zone extends 25 miles southwest across bedrock uplands (Weller 1940) and continues farther beneath the sediments of the Mississippi Embayment (Kolata et al. 1981). A northeastern extension of the zone is called the Herod Fault Zone. The Lusk Creek Fault Zone is part of the Fluorspar Area Fault Complex and contains the northwesternmost limit of known commercial fluorspar deposits.

As mapped in the Eddyville Quadrangle, the Lusk Creek Fault Zone is composed of two to three parallel

faults forming a zone 500 to 1,500 feet wide. The faults strike about N40°E, and their overall displacement is down to the southeast. Rocks of the Abbott Formation southeast of the fault zone are juxtaposed with Caseyville and uppermost Chesterian strata northwest of the zone. The downthrown block has been called the Dixon Springs Graben (Weller and Krey 1939); its southeastern margin is formed by faults that lie outside the study area.

The Caseyville Formation and uppermost Chesterian rocks northwest of the Lusk Creek Fault Zone are horizontal or dip gently within a few hundred feet of the fault zone. Close to the fault, they are sharply tilted toward the southeast and dip as steeply as 55° adjacent to the fault zone. A narrow hogback ridge of Caseyville sandstone and conglomerate follows the northwest edge of the fault zone, extending southwest from the center of Section 35, T11S, R6E. A deeply dissected plateau capped by sandstone of the middle part of the Abbott Formation occurs southeast of the fault zone in the Dixon Springs Graben. The strata in the graben dip gently northwest, with an average inclination of about 5° in the Eddyville and adjacent Waltersburg Quadrangles. Close to the fault zone, the dip gradually flattens out, then abruptly reverses. Abbott sandstone dips as steeply as 65°SE adjacent to the faults. The rocks in the graben are only slightly fractured, except immediately southeast of the fault zone where they are moderately fractured.

Within the fault zone is a central slice of highly fractured to shattered and mineralized Caseyville and undifferentiated Chesterian strata. These rocks strike northwest of the fault zone and dip 25° to 55°SE. The central slice is internally faulted, although few details can be mapped from available exposures.

The faults dip steeply southeast. Attitudes of mineralized fault planes in three fluorspar mines were listed by Weller et al. (1952, table 4):

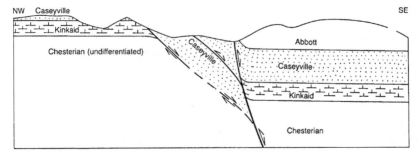

Figure 38 Profile of Lusk Creek Fault Zone in Section 31, T11S, R6E (see fig. 37 for line of section). Outcrop of Caseyville Formation is repeated in two fault slices. The middle fault is a reverse fault, as it places basal Caseyville above upper Caseyville. The other faults, and the net displacement, are normal.

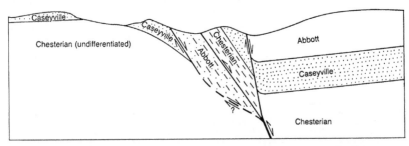

Figure 39 Profile of Lusk Creek Fault Zone in SW, Section 35, T11S, R6E (see fig. 37 for line of section). The fault zone contains two slices. The southeastern slice, containing Chesterian and Caseyville strata, is upthrown relative to rocks on either side. The northwestern slice of Abbott Formation is downdropped relative to rocks on both sides. Net displacement and final motion was normal, down to the southeast.

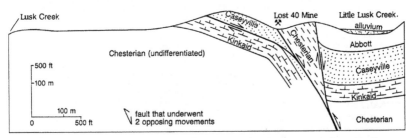

Figure 40 Profile of Lusk Creek Fault Zone just southwest of Lost 40 Mine (see fig. 37 for line of section). Wedge-shaped slice of Chesterian undifferentiated at mine is upthrown, relative to rocks on both sides. Northwest of the mine a low-angle normal fault separates Caseyville (above) from Kinkaid (below). This fault apparently follows bedding planes near the northwest brow of the hill, then cuts across bedding as the bedding steepens toward the southeast.

Lost 40 Mine
  NE NE SE, 3-12S-6E, Eddyville
  Attitude: N20°E/38-73°SE
Rock Candy Mountain Mine
  NE SE SW, 25-11S-6E, Herod
  Attitude: N40°E/85°SE
Clay Diggings Mine
  NE SE, 16-12S-6E, Waltersburg
  Attitude: N15°E/85°SE

The fault at the Rock Candy Mountain Mine is exposed in an open cut. Steep southeast dips of faults are confirmed by proprietary test drilling near the Lost 40 and Clay Diggings Mines and by a proprietary seismic profile a few miles south of the Eddyville Quadrangle. Dominantly dipslip motion is indicated by striae and corrugations on fault surfaces exposed at the Rock Candy Mountain and Clay Diggings Mines. Attitudes of drag folds and jointing also indicate dip slip. Drag-fold axes are horizontal or nearly so and strike parallel to the faults; the dominant joint set is parallel to the faults and the secondary set is perpendicular.

The Lusk Creek Fault Zone contains normal and reverse faults in and near the Eddyville Quadrangle. The southeasternmost fault, which juxtaposes Abbott Formation in the Dixon Springs Graben with older rocks in the central slice, is a normal fault throughout its length. The fault northwest of the slice of undifferentiated Chesterian rock is a reverse fault; the central slice is raised relative to Pennsylvanian rocks on either side. In NE, Section 35, T11S, R6E, the central of three faults is reverse, and the Caseyville Formation is repeated in two fault slices (fig. 38). Upraised central slices also exist at the Rock Candy Mountain Mine (Baxter et al. 1967) and at the Clay Diggings Mine in the Waltersburg Quadrangle (Weibel et al., in preparation). At the latter site, the central slice contains Ste. Genevieve Limestone, upthrown 800 to 1,500 feet relative to rocks outside the fault zone.

A narrow sliver of Abbott Formation is downfaulted between Caseyville Formation and Chesterian strata about 1/2 mile northeast of the Lost 40 Mine (fig. 39). The Abbott outcrops consist of iron-rich micaceous sandstone, carbonaceous shale, and coal, all steeply dipping.

Southwest of the Lost 40 Mine, a low-angle normal fault apparently is present on the northwest side of the fault zone. Kinkaid Limestone in the bank of Lusk Creek strikes N35°E and dips 50°SE. Directly above near the top of the bluff, Caseyville sandstone lies at nearly the same attitude. Shattered, slickensided, and mineralized sandstone and limestone float were found about midway on the hillside between the two outcrops. Northward on the hillside, Pennsylvanian sandstone directly overlies Kinkaid Limestone and both dip about 20°SE. A fault that becomes parallel to bedding northward is believed to separate Caseyville from Kinkaid at this site (fig. 40).

Geophysical findings A seismic reflection profile across the Lusk Creek Fault Zone indicates that the fault zone penetrates Precambrian crystalline basement (fig. 3). The section between the base of the Knox Group (upper Cambrian) and the inferred top of Precambrian is thicker southeast of the fault zone than to the northwest, indicating that the fault zone underwent normal movement during late Precambrian to middle Cambrian time. The southeast side dropped downward, allowing thicker sediments to accumulate there.

The Lusk Creek Fault Zone lines up with the northwest margin of the Reelfoot Rift (Ervin and McGinnis 1975), which strikes northeast and underlies the Mississippi Embayment from northeastern Arkansas to southernmost Illinois. The rift has been defined by reflection and refraction seismic studies, gravity and magnetic surveys, and limited deep drilling. It is bounded by large normal faults that penetrate crystalline basement. Pre-Knox layered rocks thicken dramatically on the downthrown sides of faults, indicating that faulting was active during pre-Knox sedimentation (McKeown and Pakiser 1982, Gori and Hays 1984, Howe and Thompson 1984).

The seismic profile (fig. 3) further shows that the southeastern boundary of the Dixon Springs Graben, the Raum Fault Zone, dips northwest and intersects the Lusk Creek Fault Zone at about the base of the Knox. Thus, the Raum Fault Zone is an antithetic fault zone, whereas the Lusk Creek Fault Zone is the master structure.

Interpretation The Lusk Creek Fault Zone evidently originated as a normal fault during latest Precambrian-early Cambrian crustal exten-

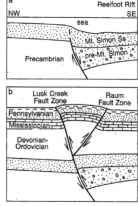

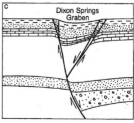

Figure 41 Interpreted history of Lusk Creek Fault Zone. (a) Reelfoot Rift forms in late Precambrian to early Cambrian time; Lusk Creek fault develops as a listric normal fault down to the southeast. Rift-fill sediment is deposited on the downthrown block. In late Cambrian time, the sea transgresses the region and Mt. Simon Sandstone is deposited on both sides of the fault. (b) In late Paleozoic time, compression from the southeast reactivates the Lusk Creek Fault Zone as a reverse fault. Raum Fault Zone develops as an antithetic reverse fault, and the block between the two fault zones is raised. (c) During the Mesozoic Era, extension takes place, allowing normal movement along Lusk Creek and Raum fault zones; the downdipped block between is the Dixon Springs graben. Note drag along fault zones and narrow slices of rock left high within the fault zones.

sion that produced the Reelfoot Rift (fig. 41a). Slippage probably ended by late Cambrian time, since Knox and younger strata do not appreciably change thickness across the fault zone. The faults that currently displace bedrock at the surface in the

63

Eddyville Quadrangle resulted from post-Atokan (Abbott Formation) reactivation of the original Cambrian fault zone.

The presence of parallel high-angle reverse and normal movements in the same fault zone seems to require two episodes of movement, compression and extension. The sharp drag and tilting of strata down to the southeast throughout the fault zone indicate that the last important movement was normal faulting, with the southeast side dropped down. The episode of reverse faulting therefore must have occurred earlier, suggesting that the region was first subjected to compression from the southeast, which reactivated the buried rift-boundary fault as a reverse fault with its southeast side upthrown (fig. 41b).

Displacement along the Lusk Creek Fault Zone during compression was probably a few hundred feet in the Eddyville Quadrangle and at least 800 feet at the Clay Diggings Mine. The Raum Fault Zone may have formed as a backthrust. The compression episode was probably induced by the Alleghenian Orogeny in late Pennsylvanian through Permian time. Later, southeastern Illinois underwent northwest-southeast extension, which created the numerous northeast-trending normal faults of the Fluorspar Area Fault Complex. The hanging wall of the Lusk Creek Fault Zone collapsed, forming the Dixon Springs Graben (fig. 41c). Some backslippage may have taken place along reverse faults, but most normal movement occurred on a newly formed fault, which presently forms the southeast margin of the Lusk Creek Fault Zone. This normal fault is believed to intersect the reverse fault at depth (figs. 38 to 40).

Displacement during extension was substantially greater than that during compression, since the net throw of the Lusk Creek Fault Zone presently is down to the southeast. The total normal displacement increases southwestward from roughly 300 feet at the eastern edge of the Eddyville Quadrangle to 1,500 feet at Clay Diggings; these are minimum figures, since the central slices may have been higher before normal faulting. Drag during normal movement rotated bedding within and near the fault zone, so that it dips southeast. Any drag folds formed during reverse faulting were obliterated. The

tilted and downthrown slices along the northwestern side of the fault zone (figs. 38 to 40) are entirely products of the extensional episode.

The normal faults probably are of early Mesozoic age. Numerous grabens developed in the eastern United States during the Triassic Period; the Atlantic Ocean began to open during the Jurassic. Subsidence of the Mississippi Embayment began in the Cretaceous Period. The Lusk Creek Fault Zone apparently does not deform upper Cretaceous strata of the Embayment in southernmost Illinois (Kolata et al. 1981).

This scenario of reverse faulting followed by normal faulting is the same as that proposed for the Rough Creek-Shawneetown Fault System (Smith and Palmer 1981, Nelson and Lumm 1984, Kolata and Nelson, in press). The Rough Creek-Shawneetown, like the Lusk Creek, contains parallel high-angle normal and reverse faults that outline narrow slices of older rock apparently sheared off the hanging wall (south or southeast block) during normal movements. Both fault systems are reactivated late Precambrian-early Cambrian rift-boundary faults. They form a continuous shear zone and are products of the same stresses.

McCormick Anticline

Name and extent Brokaw (1916) mapped an anticline in northwestern Pope County, Illinois, and named it the McCormick Anticline after the village of McCormick in the Stonefort Quadrangle. Jacobson and Trask (1983), on the basis of preliminary mapping, thought that faulting dominated over folding and suggested changing the name to McCormick Fault Zone. Continued mapping reveals an anticline with subordinate monoclines, synclines, and faults. The name McCormick Anticline is retained for the sake of simplicity; the structure could be called an anticlinorium, a faulted anticline, or an anticline and fault zone.

The McCormick Anticline begins just east of the Eddyville Quadrangle, at the western edge of the Shawneetown Fault Zone, and extends westward across the north-central part of the Eddyville Quadrangle, gradually curving toward the southwest (figs. 2 and 37). In the Stonefort Quadrangle, it turns to a heading of about S50°W, passing just

north of McCormick and exiting at the southwestern corner of the quadrangle. It continues across the Bloomfield Quadrangle, gradually turning more southward and dying out in southern Johnson County, east of Vienna (Weller 1940, Nelson in prep.).

Structure The McCormick Anticline is an asymmetrical fold, with a steep and narrow northwest limb and a broad and gentle southeast limb. Dips on the northwest flank range from 12° to 45° and typically are 20° to 35°. The southeast limb dips less than 10° in most places, but locally is steeper, especially near faults. Structural relief from the anticlinal crest to the trough of the Battle Ford Syncline, northwest of the anticline, is at least 500 feet in most places and reaches a maximum of about 800 feet in the eastern part of the Stonefort Quadrangle. The relief is considerably less, 200 to 300 feet, from the anticline to the Bay Creek Syncline on the southeast.

Superimposed upon the larger fold are several smaller anticlines (see fig. 37 and structure-contour lines on geologic maps). Three of these in the Eddyville Quadrangle have curved axes. The eastern ends of the minor anticlinal axes strike east-west, parallel to the larger structure; westward they curve toward the southwest. Thus, they form a right-handed en echelon pattern. Subordinate anticlines in the Stonefort Quadrangle have slightly sinuous axes nearly parallel with the larger fold, and form a relay rather than an en echelon pattern.

In profile, the folds vary from smoothly arched, to boxlike, to sharp crested. Smooth-crested folds are exemplified near Burden Falls in the Stonefort Quadrangle (Sections 10, 11, and 15, T11S, R5E) and north of Pleasant Ridge Church (Sections 6 and 7, T11S, R6E) in the Eddyville Quadrangle. A box fold with a slightly tilted crest roughly 1/2 mile wide occurs in the eastern part of the Eddyville Quadrangle (fig. 42). Sharp anticlines, faulted along their crests, are found southwest of McCormick in the Stonefort Quadrangle. The transition from steeply dipping beds on the northwest limb to gently dipping beds farther northwest is abrupt in most places. The dip changes from 45° to less than 10° in a few yards near the southwest corner of the Stonefort Quadrangle.

64

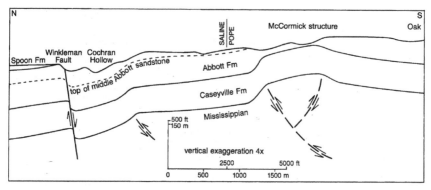

Figure 42 Cross section of McCormick structure and Winkleman Fault in the northeastern part of the Eddyville Quadrangle (see figure 37 for line of section). The McCormick structure is an asymmetrical box-fold. Sharp flexures here and on the south side of Cochran Hollow may be faulted at depth. Winkleman Fault apparently is a normal fault dipping steeply southeast; it represents the eastern terminus of the New Burnside structure.

Faults have been mapped along most of the length of the McCormick Anticline (fig. 37). Some faults pierce the crests of sharp anticlines. Others diagonally cross the southeast flank of the structure. Both categories of faults in map view are rotated slightly counterclockwise to the trend of the larger structure, so they describe a right-handed en echelon pattern. This right-handed arrangement is the same as that noted for the three small anticlines in the Eddyville Quadrangle. Few faults have been mapped on the northwest flank of the structure, where dips are steepest and abrupt changes in dip are common.

Most of the faults that have been observed are normal faults that dip 60° or steeper. A normal fault dipping 45°NW with about 25 feet of throw is well exposed along the ravine north of Pleasant Ridge Church, near the center of N½, Section 7, T11S, R6E (Eddyville field stations 287 and 288). Smaller normal faults that dip 45° to 80° are exposed along an adjacent cliff of Pounds Sandstone. A fault exposed in a stream bed in SE NW SE, Section 36, T11S, R5E (Stonefort field station 357) strikes N20°E and dips 70°SE. Two high-angle reverse faults and several steeply dipping normal faults are exposed along Ogden Branch in SW NE SW, Section 16, T11S, R5E (Stonefort field station 656).

On other faults not completely exposed, nearly straight mapped traces across rugged terrain and

attitudes of adjacent rocks indicate steep dips.

The few slickensides that have been noted indicate dominantly dip-slip displacement, as do the orientations of drag-fold axes, nearly horizontal and parallel to faults.

On some faults, the direction of last movement deduced from drag folds and tilt of strata is opposite to the stratigraphic displacement. One example from the southwestern part of the Stonefort Quadrangle is illustrated in figure 43. Strata dip steeply southeast along both sides of this fault, suggesting a large downthrow

to the southeast; yet the stratigraphic displacement is down to the northwest. Two other cases are shown (figs. 44 and 45) in which two nearly parallel faults outline a steeply tilted central slice of Caseyville sandstone. In the examples in figures 44 and 45 and elsewhere, the last movement was evidently normal and down to the southeast, partially or completely canceling an earlier movement in which the southeast block was raised. The geometry is nearly identical to that of the Lusk Creek Fault System and implies the same sequence of structural events.

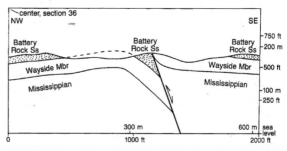

Figure 43 Cross section through a fault on the McCormick Anticline southwest of the Zion Church, Stonefort Quadrangle (section runs S45°E from center of Section 36, T11S, R4E; see fig. 37 for line of section). The fault strikes approximately N20°E and dips 70° southeast, as exposed in stream bed near line of section. The stratigraphic displacement is down to the west; however, the strata on both sides of the fault dip southeast, indicating that the last movement on the fault was down to the southeast (normal) following an earlier episode of reverse faulting in which the southeast side was raised.

65

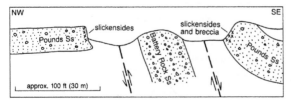

Figure 44 Field sketch of relationships in NW SE SW, Section 5, T11S, R6E, Eddyville Quadrangle (see fig. 37 for line of section). Steeply tilted, conglomeratic sandstone is found between two outcrops of Pounds Sandstone, with indications of faults on both sides. Assuming that the faults dip steeply southeast (as suggested by the tilt of the strata), we have an upthrown central slice but almost no displacement across the fault zone. Drag folding of sandstone southeast of the fault zone indicates that the last movement was down to the southeast. This configuration is nearly identical to that of the Lusk Creek Fault Zone.

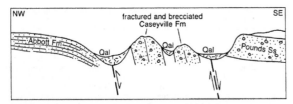

Figure 45 Cross sectional sketch, not to scale, of field relationships near head of ravine just west of McCormick, NE SE NE, Section 20, T11S, R5E (see fig. 37 for line of section). Outcrops of steeply dipping to vertical, fractured, and brecciated Caseyville sandstone are found between gently dipping Abbott sandstone on the northwest. This configuration suggests two periods of movement, in which the Caseyville strata southeast of the fault zone were first raised and then lowered. The southeasterly tilt and the drag folds were imparted during the second movement, which involved normal faulting.

Small trace-slip faults have been observed in several places along the north or northwest flank of the McCormick Anticline. (As defined by Beckwith (1941), a trace-slip fault is perpendicular or oblique to bedding and underwent slip parallel to bedding, so that no offset of bedding is apparent. A strike-slip fault in horizontal strata is one example of a trace-slip fault.) The strike of these faults varies from N15°E to N70°W, making them oblique or nearly perpendicular to the large mapped faults. The dip is normal to bedding, and the faults bear prominent grooves and striations parallel to the trace of bedding (thus the term trace-slip fault). In no case could the direction or amount of offset be determined. Because all observed trace-slip faults are confined to single outcrops and contain no noticeable gouge or breccia, their extent and displacement are believed to be small. One of the best places to see

this type of fault is on Pounds Sandstone just east of the north-trending stream 600 feet from the south line, 2,100 feet from the east line of Section 6, T11S, R6E (Eddyville field station 285).

Joints Sets of parallel planar fractures, normal or very steeply inclined to bedding, are conspicuous in many outcrops along and near the McCormick structure. Such joints are seen in many lithologies, being best developed in well-indurated, thin-bedded, quartzose sandstones and siltstones. In these rocks, joints commonly are spaced a few inches apart and locally are a fraction of an inch apart. Massive sandstones typically contain very large planar fractures spaced several feet to tens of feet apart, but near faults the sandstones are closely jointed. Orientations of joints are summarized in the form of rose diagrams (fig. 46). Although the sampling was not statistically rigor-

ous, enough measurements were taken to expose some definite trends.

Along the McCormick Anticline in the Eddyville Quadrangle (fig. 46a), the joint directions cluster in two well-defined maxima. The stronger is N50°-60°E, which is parallel to the main direction of faults. The secondary maximum is N50°-60°W, nearly perpendicular to the first set. Joints are most numerous and closely spaced near faults. Well-developed joints also are common along the crests and steep north flanks of the folds. Joint data from the Stonefort Quadrangle (fig. 46b) show considerably more scatter and represent fewer readings than in the Eddyville Quadrangles. Northeast and east-west trends of joints seem to be preferred. Northeast-trending joints seem to dominate northeast of McCormick village, whereas east-west joints are most common southwest of McCormick. The rose diagram for joints south of the McCormick Anticline in the eastern part of the Eddyville Quadrangle (fig. 46c) displays two strong, nearly equal maxima, one at N10°-15°E and the other at N80°-90°W. The first set is parallel to the Shawneetown Fault Zone, and the second set is perpendicular to the Shawneetown and parallel to the McCormick Anticline. Farther south (not shown in the rose diagram), the joint orientation changes to NE-SW and NW-SE, parallel and normal to the Lusk Creek Fault Zone.

Joints are rare and poorly developed south of the McCormick Anticline in the Stonefort Quadrangle and western half of the Eddyville Quadrangle. Jointing also is inconspicuous in the Battle Ford Syncline, north of the McCormick Anticline. Measurements were recorded at only 15 field stations within the syncline. At these localities, a preferred orientation of N45°-60°E was observed.

New Burnside Anticline

Name and extent The New Burnside Anticline was first described and named by Brokaw (1916). Weller (1940) renamed the eastern part of the structure the Stonefort Anticline, which he considered to be separate from the New Burnside. Jacobson and Trask (1983) thought the structure was dominated by faulting and proposed the name New Burnside Fault Zone. Mapping for this report shows that the New Burnside and Stonefort segments are intimately related, although separated by a saddle, and that faulting is subordinate to folding. We refer to both by the older name, New Burnside Anticline.

The New Burnside Anticline is north of and trends roughly parallel to the McCormick Anticline (figs. 2 and 37). The axis of the New Burnside extends westward across the northern edge of the Eddyville Quadrangle and curves to trend west-southwest across the$_e$ Stonefort Quadrangle. In the Creal Springs Quadrangle, it turns southwestward, passing just south of New Burnside and Parker. Beyond Parker the structure gradually broadens and flattens; it dies out northwest of Tunnel Hill. A series of northeast-trending normal faults associated with the New Burnside extends into the southern parts of the Harrisburg and Rudement Quadrangles north and northeast of the current study area. The New Burnside and McCormick Anticlines both are arcuate in map view, with the convex side facing northwest.

Structure The New Burnside Anticline is geometrically similar to the McCormick Anticline, but has less relief and is less faulted (fig. 37 and geologic maps). The steeper north flank dips 10° to 25° in most places and locally is as steep as 60°. Dips on the southeast flank generally are less than 10°. Both flanks flatten as the anticline plunges southwestward in the Creal Springs Quadrangle. The three areas of maximum relief are near the northwest corner of the Eddyville Quadrangle, north of Sand Hill in the Stonefort Quadrangle (Section 4, T11S, R5E), and near the border of the Stonefort and Creal Springs Quadrangles. In these

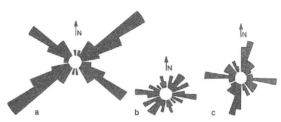

a   b   c

Figure 46 Rose diagrams of joint orientations along and south of McCormick Anticline. Average readings from (a) 43 sites along the McCormick Anticline in the Eddyville Quadrangle; (b) 23 sites along the McCormick Anticline along the Stonefort Quadrangle; and (c) 23 sites south of the McCormick Anticline in the Eddyville Quadrangle.

places, the elevation of the top of the Abbott Formation drops 400 feet in 1/2 to 3/4 mile north and 200 to 250 feet in 1 mile south of the anticlinal crest. The relief is only about 130 feet in the structural and topographic saddle at Oldtown, Stonefort Quadrangle. Railroad cuts expose the structure well and show it to be a slightly faulted monocline (fig. 19).

The anticline plunges abruptly at its east end in NE NE, Section 34, T10S, R6E, Eddyville Quadrangle. A northeast-facing monocline of low relief and a southeast-dipping fault branch off the east end of the anticline. At its southwestern end, the New Burnside Anticline merges almost imperceptibly into a region of homoclinal northerly dip.

East of Oldtown, the New Burnside consists of a series of anticlines, arranged in subparallel relay or right-handed en echelon pattern. Near the eastern end of the structure, some of the anticlines are cuspate in profile. This is best seen on the steep west bluff of Blackman Creek in SE SW SE, Section 27, T10S, R6E (Eddyville field station 85). Dips on both flanks of the fold steepen toward the crest,

which is a sharp but apparently unfaulted hinge (fig. 47). Roadcuts on Route 145, a short distance west of Blackman Creek, show a sharply hinged box fold with the south limb faulted. Farther west, along Battle Ford Creek, two parallel cuspate anticlines, separated by a shallow syncline, have been mapped (fig. 48).

The anticline near Murray Bluff is sharp crested; the southern limb is cut off by a fault that strikes parallel (N50°E) to the fold axis (fig. 49). The fold and fault terminate abruptly near Bill Hill Hollow, and another anticline arises 2,000 feet north near Stonefort Bluff. The sudden offset of the fold axis and the linear trend of Bill Hill Hollow suggest a tear fault along the hollow. Such a fault, if it exists, is entirely concealed by alluvium. West of Stonefort Bluff, the axis of the New Burnside Anticline is sinuous and possesses saddles and elongate domes, but it is a continuous fold. In profile, it is smoothly arched, not cuspate or boxlike.

Faulting along the New Burnside Anticline is most pronounced in the Eddyville Quadrangle. The trace of the Winkleman Fault (Cady 1926)

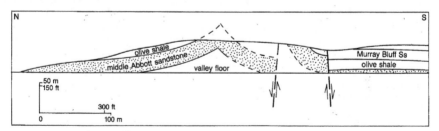

Figure 47 Profile of New Burnside structure on the west side of Blackman Creek (see fig. 37 for line of section). Note sharp-crested anticline; dips increase toward crest on both flanks.

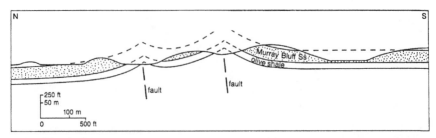

Figure 48 Profile of New Burnside structure where it crosses Battle Ford Creek, in the north-central part of the Eddyville Quadrangle (see fig. 37 for line of section). Two sharp-crested anticlines, separated by a shallow syncline, are present. The northern anticline changes into a fault eastward, whereas the southern anticline changes into a fault eastward and dies out westward.

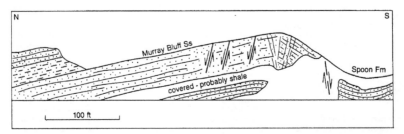

Figure 49 Profile of a hillside east of a ravine, near center E 1/2, Section 25, T10S, R5E, Eddyville Quadrangle (field station 29). Note sharp fold hinge in Murray Bluff Sandstone; dip changes within a few feet from 10°N to 35°-40°S. Many fractures but no offsets are visible in the hinge; small normal faults are exposed on ledges north of the hinge. East of the fold hinge a concealed fault juxtaposes sandstone of lower Spoon Formation with Murray Bluff.

trends N60°E and extends from the northeast corner of the quadrangle to the eastern end of the New Burnside Anticline, where it curves west and parallels the fold for about 1/2 mile. Then it again bends sharply to the southwest and connects with a complex zone of faults along the McCormick Anticline. The Winkleman Fault is a high-angle normal fault with a maximum throw of about 100 feet.

An east-west trending fault in S1/2, Section 29, T10S, R6E, is exposed at several places in a stream bed and appears to dip almost vertically. It passes laterally at both ends to unfaulted sharp-crested anticlines. In the northwestern part of the Eddyville Quadrangle near Murray Bluff is a large high-angle fault that strikes N50°E; its southeast side is downthrown as much as 200 feet. The fault lies a few tens of feet southeast of a sharp-crested anticline. At the north edge of the map

area and in the adjacent Harrisburg Quadrangle, a series of northeast-trending faults outlines a graben. Several of these faults are exposed in highwalls in the Brown Brothers Company No. 1 Mine and were exposed before reclamation in the No. 2 Mine farther north. All are normal faults that dip 70° or steeper and bear vertical striae and corrugations indicative of pure dip-slip displacement (fig. 50). Displacements of faults exposed in the coal mines range from a few feet to 150 feet.

Faults along the anticline in the Stonefort and Creal Springs Quadrangles are more widely spaced and have smaller displacements than those in the Eddyville Quadrangle. The railroad cut south of Oldtown reveals numerous high-angle normal faults and a few high-angle reverse faults that strike N75°W to N65°E, and have displacements ranging from a few inches to 10 feet. Five northeast-trending en echelon faults

have been mapped along the anticline in the western parts of the Stonefort and Creal Springs Quadrangles. All of them dip steeply and have throws of a few tens of feet. They extend farther south than north of the fold axis; several faults nearly reach the trough of the Battle Ford Syncline. The four eastern faults probably are simple normal faults. The westernmost fault has drag inconsistent with stratigraphic offset. A sandstone bed appears to be downthrown about 20 feet to the northwest, but rocks dip steeply southeast on both sides of the fault plane in SW NW NW, Section 25, T11S, R3E. This suggests that the southeast block was first raised and then lowered, as on the Lusk Creek Fault Zone and several faults in the western part of the McCormick Anticline.

A high-angle reverse fault that strikes N65°W was observed in a stream cut north of the anticline in

Figure 50 (a) Fault surface exposed by mining in reclaimed Brown Brothers Excavating No. 2 Mine just north of the Eddyville Quadrangle. The fault plane dips steeply, exhibiting vertical slickensides, normal movement, and little folding of strata. (b) Sandstone breccia on fault surface on highwall of the abandoned Brown Brothers No. 1 Mine just north of the Eddyville Quadrangle (SE NW, Section 20, T10S, R6E, Harrisburg Quadrangle).

N1/2 NE NW, Section 6, T11S, R5E (Stonefort field station 747). The attitude and location of this fault are anomalous for faults along the New Burnside Anticline.

Trace-slip faults similar to those described along the McCormick Anticline exist along the New Burnside also. Several prominent ones (fig. 51) were observed on a sandstone ledge in SW SE NW, Section 35, T10S, R5E (Stonefort field station 558). The site is on the northwest flank of the anticline where bedding dips 25° to 30° northwest. The faults are nearly perpendicular to the anticline and dip vertically. Other trace-slip faults can be seen in the railroad cut about 1,000 feet south of the grade crossing at Oldtown and in a railroad cut south of Parker in the Creal Springs Quadrangle in NE SW NW, Section 17, T11S, R4E. These faults also strike northwesterly.

Joints Rose diagrams of joints along the New Burnside Anticline (fig. 52) represent average readings at a total of 103 sites in all three quadrangles. A preferred northeasterly orientation of joints is evident in all three quadrangles. More precisely, the principal trend is N50°-60°E in the Eddyville Quadrangle and N40°-50°E in the Stonefort and Creal Springs Quadrangles. Thus, joints are generally subparallel with nearby normal faults along the anticline. A secondary northwest-trending set of joints, indicated on the rose diagrams, is much weaker than the northeast-trending set.

A rose diagram (fig. 52e) of 37 joint measurements northwest of the New Burnside structure in the Creal Springs Quadrangle reveals no definite preferred orientation. A diagram (fig. 52d) for the area southeast of the structure, representing 66 stations, shows a strong peak at N10°-20°E and a secondary peak at N50°-60°E.

Battle Ford Syncline

The Battle Ford Syncline, named in this report, lies between the McCormick and New Burnside Anticlines. Its name is taken from Battle Ford Creek, which follows the synclinal axis for a short distance in the north-central Eddyville Quadrangle. The syncline extends about 18 miles from W1/2, Section 34, T10S, R6E, Eddyville Quadrangle, to the headwaters of Sugar Creek in Section 35, T11S, R3E, Creal Springs Quadrangle (fig. 37). The Battle Ford Syncline is a topographic depression as well as a

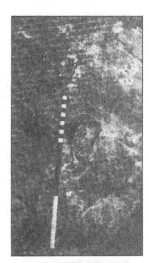

Figure 51 Trace-slip fault near center of W1/2, Section 35, T10S, R5E, Stonefort Quadrangle. This vertical fault strikes N45°W. Striae and corrugations on the fault surface plunge about 30° northwest, parallel to the trace of the bedding on the fault surface. The staff is graduated in feet and inches.

69

structural one. The upper reaches of Battle Ford Creek, the Little Saline River, and Sugar Creek all follow the structural trough. Upland surfaces within the syncline are largely dip slopes on the Murray Bluff Sandstone Member of the Abbott Formation.

The Battle Ford Syncline is broad and asymmetrical. Its width increases from a little over 1 mile on the east to about 4 miles on the west. The trough or axis is sinuous and lies generally within 1 mile of the crest of the New Burnside Anticline. Relief from synclinal trough to New Burnside crest varies from less than 50 feet to about 300 feet. The south flank of the syncline dips at an average rate of less than 1° until close to the McCormick Anticline, where the inclination abruptly steepens. Structural relief between the trough of the syncline to the crest of the McCormick Anticline ranges from 400 feet to more than 750 feet (see structure contours on geologic maps).

At its western end, the Battle Ford Syncline loses its identity where the New Burnside Anticline fades away. At the eastern end, the syncline is tightly closed off by a west-facing flexure, roughly parallel to Blackman Hollow. East of this point, the strata slope steadily northward, interrupted only by the Winkleman Fault. The eastern end of the syncline is cut by the Winkleman Fault, which diagonally links the McCormick and New Burnside Anticlines. Displacement on this fault is approximately 125 feet down to the southeast. The downdropped block southeast of the Winkleman Fault contains the structural lowest point (structurally) in the Battle Ford Syncline. This roughly triangular basin is bordered by the fault, the McCormick Anticline, and the eastern closure of the syncline. The drainage in this little basin gathers inward from three directions, flowing down sandstone dip slopes and merging at the fault to a single northwest-flowing stream deeply incised in Murray Bluff Sandstone northwest of the fault.

Bay Creek Syncline

The name Bay Creek Syncline is applied in this publication to an enclosed structural depression that lies southeast of the McCormick Anticline. It is named for Bay Creek, which approximately follows the axis of the syncline from its source to Bell

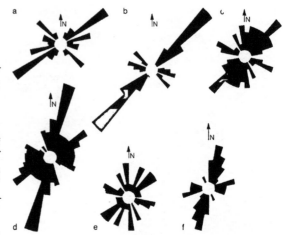

Figure 52 Rose diagrams of joint orientations on and near New Burnside Anticline in the Eddyville Quadrangle. Average readings from (a) 27 sites along the New Burnside Anticline, Eddyville Quadrangle; (b) 23 sites along the New Burnside Anticline, Stonefort Quadrangle; (c) 53 sites along the New Burnside Anticline, Creal Springs Quadrangle; (d) 66 sites southeast of the New Burnside Anticline, Creal Springs Quadrangle; (e) 37 sites northwest of the New Burnside Anticline, Creal Springs Quadrangle; and (f) 33 sites along Little Cache Fault Zone and vicinity, Creal Springs Quadrangle.

Smith Springs. The syncline extends about 12 miles southwest from the east-central part of the Eddyville Quadrangle, through Watkins Ford, to the center of the southern border of the Stonefort Quadrangle, where it leaves the study area.

The trough of the Bay Creek Syncline lies roughly 2 miles southeast of the McCormick Anticline and strikes parallel to the latter. The dips on both flanks average 1.5° and are quite regular, in common with the southeast flank of the McCormick Anticline. The northwest flank, in common with the southeast flank of the McCormick Anticline, curves from northeast to east. Structural contour lines on the south flank trend slightly north of due east. The trough of the syncline is 300 to 400 feet lower in elevation than the crest of the anticline (see structure contours on geologic maps). The eastern closure of the Bay Creek Syncline is gradual; the western closure has not been mapped.

Geophysical Expression of Anticlines

Geophysical data available for the study area include several regional magnetic and gravity surveys, one published local magnetic survey, un-

published and preliminary gravity and magnetic surveys of the Stonefort and Creal Springs Quadrangles, and two proprietary seismic reflection profiles.

Regional maps of the gravity field in southern Illinois include those of McGinnis et al. (1976) and Hildenbrand et al. (1977). Regional magnetic maps were made by McGinnis and Heigold (1961), Lidiak and Zietz (1976), and Johnson et al. (1980). All were based on control points spaced 1 mile or more apart. They show regional trends but do not indicate anomalies that can be related to specific structures mapped in the study area.

A ground magnetic survey by Heigold and Robare (1978) covered most of the present study area, including all of Johnson County, southwestern Saline, and northwestern Pope County. Three detailed profiles were made; the only one in our study area crossed the McCormick Anticline along the Illinois Central Railroad in the Stonefort Quadrangle. This profile was considered inconclusive by its authors. Heigold and Robare's map of residual magnetic intensity (fig. 53) indicates an overall magnetic

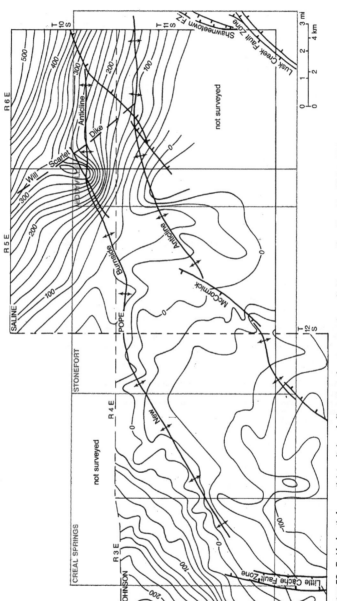

Figure 53 Residual vertical magnetic intensity in and adjacent to study area; contour interval is 20 gammas (from Heigold and Robare 1978).

71

gradient rising toward the northwest across the study area. The contour lines of magnetic intensity form a slightly arcuate pattern roughly paralleling the McCormick and New Burnside Anticlines. Considering that sedimentary rocks generally have very low magnetic susceptibility, Heigold and Robare believed the magnetic gradient to reflect configuration of the Precambrian surface. They regarded the McCormick and New Burnside Anticlines as products of compression from the southeast. Although they saw no indication that the McCormick and New Burnside Anticlines affect basement, Heigold and Robare speculated that the arcuate form of these structures may have been controlled by the configuration of the basement surface.

A suite of gravity and magnetic maps of the Creal Springs Quadrangle was prepared by Lawrence Malinconico of Southern Illinois University, Carbondale. Malinconico and Steven Brennecke also produced a map of the first-order residual magnetic field in the Creal Springs Quadrangle. These maps, like Heigold and Robare's map, show magnetic field increasing toward the northwest. Several oval to irregular shaped magnetic highs and lows, 1/2 to 2 miles across at the long axis, also are shown on the maps. Several pronounced lows are present just south of the crest of the New Burnside Anticline. Malinconico (personal communication, 1988) suggested that these could reflect thickened sedimentary section, possibly due to overthrusting beneath the south flank of the anticline. The Bouger

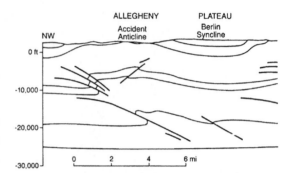

Figure 54 Cross section in western Maryland and West Virginia, from Kulander and Dean (1986). Narrow, steep-flanked anticlines and broad, intervening synclines northwest of the Allegheny structural front are related to imbricate thrust faults above major detachment surfaces in Lower Cambrian and Upper Ordovician shales.

gravity map shows a strong and sinuous gravity gradient down to the north, just south of the New Burnside Anticline. This suggests possible duplication by thrusting of the relatively dense Valmeyeran carbonate section on the south limb of the fold (Malinconico, personal communication, 1988). However, no corresponding gravity and magnetic features are associated with the McCormick Anticline. Gravity and magnetic data are inconclusive on the nature of the anticlines at depth.

Two proprietary seismic reflection profiles from the study area were viewed by the authors of this report. A north-south profile in the Creal Springs Quadrangle crossed the western end of the New Burnside Anticline. The anticline was evident

and appeared to be offset by one or more faults that apparently flatten with depth. The Creal Springs profile was of poor quality, and reflectors could not be correlated with stratigraphic units. The other seismic profile (fig. 3), of much higher quality, ran north-south through the Eddyville Quadrangle, crossing the McCormick and New Burnside Anticlines and the Lusk Creek Fault Zone. This profile showed strong arching of reflectors near the surface on both anticlines and also indicated two or three south-dipping listric faults, stacked one above another beneath the crest of the McCormick Anticline. The New Burnside Anticline apparently was cut by three fairly high-angle planar faults, of which two dipped south and one dipped north.

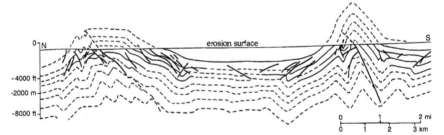

Figure 55 Profile in the Ruhr coal basin, West Germany, showing deformation of Upper Carboniferous strata as a result of the Variscan orogeny (late Paleozoic). Profiles are based on borehole records and surveys in multiple-seam underground coal mines. Upright boxlike and sharp-crested anticlines contain imbricate reverse faults, mostly high-angle, on crests and flanks. Regional direction of thrusting was south (right) to north. Seismic reflection surveys in northeastern France indicate that the Variscan thrust faults flatten at depth into decollements in Devonian strata. (Raoult and Meilliez 1987; Drozdzewski et al. 1980)

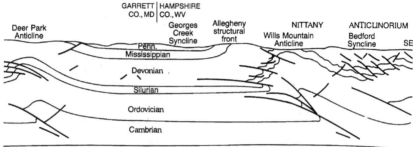

GARRETT | HAMPSHIRE
CO., MD | CO., WV

Deer Park
Anticline

Georges
Creek
Syncline

Allegheny
structural
front

NITTANY
Wills Mountain
Anticline

ANTICLINORIUM
Bedford
Syncline   SE

Penn.
Mississippian

Devonian

Silurian

Ordovician

Cambrian

Precambrian

Normal and reverse faults were interpreted. A strong deep reflector on this profile, probably at the base of the Knox Group (upper Cambrian-lower Ordovician) showed neither folding nor faulting in the area of the McCormick and New Burnside Anticlines. Thus, this profile indicates that the McCormick and New Burnside Anticlines are not rooted in basement, but are detached within the sedimentary column.

Origin of McCormick and New Burnside Anticlines

The structural geometry of the McCormick and New Burnside Anticlines, as mapped at the surface, is consistent with the conclusion from seismic evidence that the anticlines are products of detached thrusting from the southeast.

Detached thrust folding typically yields sets of elongate, parallel, arcuate anticlines and synclines, with the steeper limbs facing away from the direction of tectonic transport. Most of the world's folded mountain belts exhibit such a pattern. The McCormick and New Burnside Anticlines fit the same pattern, with obvious differences in scale and tectonic setting.

Striking similarities are seen in profile between the McCormick and New Burnside Anticlines and parts of the moderately compressed section of the Allegheny Plateau, northwest of the Allegheny structural front in West Virginia and western Maryland (fig. 54). In the Allegheny example, anticlines are bounded where northwest-verging thrust faults ramped upward from decollements located within the incompetent Martinsburg Shale (upper Ordovician) and in basal Cambrian shales (Kulander and Dean 1986, Rodgers

1963). The Deer Park and Accident Anticlines (fig. 54) are relatively narrow and have steep flanks, especially on the northwest; the adjacent Berlin and Georges Creek Synclines are broad and nearly flat bottomed. The crest of the Deer Park Anticline is smoothly arcuate, whereas that of the Accident Anticline is boxlike. Both anticlines contain imbricate (overlapping) southeast-dipping thrust faults that do not reach the surface; both also have northwest-dipping backthrusts. A paired forethrust and backthrust can produce a box fold. Note that some faults in figure 54 are entirely detached (not connected to decollements). Such faults can result from "crowding" of strata on the limbs of a fold. Note that two sets of faults, emanating from two decollements, are stacked in the Deer Park Anticline. Similar stacking of faults was observed on the McCormick-New Burnside seismic profile.

An example from the Ruhr coal basin of West Germany (fig. 55) illustrates boxlike and sharp-crested anticlines in a thrust-fold belt. The Ruhr area has undergone a greater degree of horizontal shortening than has the McCormick-New Burnside area, so folds are sharper and more tightly crowded in the former than in the latter. Nevertheless, similarities in fold geometry are apparent.

The present study area is located within the North American craton, more than 200 miles from the Alleghenian and Ouachita fold-thrust belts at the late Paleozoic continental margin. It is highly improbable that decollements traversed such a distance from these orogenic belts to southern Illinois. However, thrust-fold structures have been documented in cratonic settings and provide

direct analogues to the present case. Petersen (1983) cited several examples of detached thrust folds that are related to nearby high-angle basement-rooted reverse faults. An example from the Anadarko Basin is shown (fig. 56). Petersen described and illustrated comparable structures from the Sweetwater Uplift and Bighorn Basin in Wyoming. These anticlines occur singly or in small parallel or en echelon groups, generally several miles basinward from the parent high-angle basement-rooted faults.

I personally have mapped a well-exposed segment of such an anticline, the Beaver Creek Anticline in the eastern part of the Bighorn Basin. The anticline is about 5 miles long, averages 1/4 mile wide, and has 200 to 300 feet of structural relief. The fold hinge varies within short distances, from smoothly curved to box-like to hingelike. In map view, the fold axis is slightly to sharply sinuous and is arcuate overall, the convex side facing the basin (westward). Thrust faults, with dips ranging from horizontal to nearly vertical, cut the crest and limbs of the anticline. Except for the prominence of thrusting, this structure closely resembles the McCormick and New Burnside Anticlines in both scale and geometry. Noggle (1986), who mapped a larger area, interpreted the Beaver Creek Anticline as a detached thrust-fold connected with major basement-seated thrust faults at the west flank of the Bighorn Uplift, 2 to 3 miles distant.

The McCormick and New Burnside Anticlines, by analogy to the examples above, are here interpreted as thrust-folds linked by decollements to the basement-seated Lusk Creek Fault Zone (fig. 57). Reverse faulting on the Lusk Creek and fold-

73

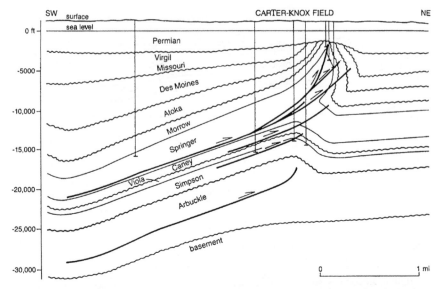

Figure 56 Example of foreland detachment structure in the Anadarko Basin, Grady and Stephens Counties, Oklahoma. This thrust/fold structure developed in response to basement uplift of the Wichita Mountains (near the end of the Pennsylvanian Period) on the southwest flank of the Anadarko Basin, along a high-angle reverse fault. Basement uplift along Lusk Creek Fault Zone induced detached thrusting along the McCormick and New Burnside Anticlines. (From Petersen 1983)

ing of the anticlines took place during the same compression episode, presumably late Pennsylvanian and Permian. Stacking of thrust faults on seismic profiles indicates at least three décollement horizons. Likely units include Chesterian shales, the New Albany Shale (mainly upper Devonian), the Maquoketa Shale (upper Ordovician), and a shaly interval at or near the base of the Knox Group, possibly correlative with the Davis Shale (upper Cambrian) of southeastern Missouri.

Most of the faults at the surface along the anticlines are normal faults that probably originated as reverse faults during compression, and later underwent normal movement. Most thrust faults, however, probably did not reach the surface. Normal faults may

have propagated upward from thrust faults at depth or developed in previously unfaulted areas where folding and jointing had weakened the rocks. The en echelon arrangement of many faults suggests that the extension stress was not quite normal to the anticline, but rotated slightly counterclockwise. Alternatively, a component of left-lateral wrenching may have been involved.

The trace-slip faults, which are mostly normal or highly oblique to the fold axes, probably originated during compression as either small tear faults or conjugate strike-slip faults. Most likely, they formed while bedding was still nearly horizontal, and subsequently rotated as the rocks were tilted.

Time of Deformation of Anticlines

I believe the major deformation on the McCormick and New Burnside Anticlines was post-Desmoinesian (Spoon Formation). A compressional event, probably related to the late Pennsylvanian-Permian Alleghenian orogeny, folded the strata. Subse-

quent extension connected with Triassic-Jurassic rifting created normal faults. Events on the detached anticlines and on the basement-rooted Lusk Creek and Rough Creek-Shawneetown Fault Zones are intimately related.

No evidence exists for pre-Pennsylvanian structural movements. Widespread deformation elsewhere in the Illinois Basin during the time period represented by the Mississippian-Pennsylvanian hiatus (Kolata and Nelson, in press) did not affect the present study area. Evidence consists of boreholes and outcrops that show Goreville and Grove Church Members of the Kinkaid at or near the crests of the anticlines. If any uplift had taken place during the pre-Pennsylvanian hiatus, these units would have been eroded.

Initial uplift of the McCormick Anticline during Caseyville sedimentation is suggested by observations that the Caseyville is thinner on the anticline than in adjacent synclines. The Drury and Wayside Members in particular appear to be thinner on the anticline than elsewhere. Lithofacies

74

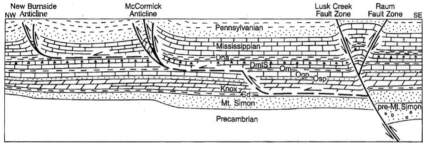

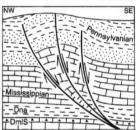

Figure 57 Hypothetical origin of McCormick and New Burnside Anticlines and associated faults. (a) In late Paleozoic time, reverse movement occurs along the Lusk Creek Fault Zone. Horizontal thrusts (decollements) propagate toward the northwest and ramp upward; imbricate thrusting produces anticlines. Likely decollement zones include (1) upper Cambrian Davis Shale, (2) upper Ordovician Maquoketa Shale, and (3) Upper Devonian New Albany Shale. (b) Mesozoic extension allows normal, down-to-the-southeast movement on imbricate faults beneath anticlines, in some cases canceling early reverse displacements.

Dna	New Albany
DmlS	Middle, Lower Devonian; Silurian
Om	Maquoketa
Ogp	Galena-Platteville
Osp	St. Peter
€d	Davis Shale

and thickness data are too sparse to support further speculation.

A series of imbricate thrust faults is exposed in Wayside strata near the south portal of the Illinois Central Railroad tunnel in NE, Section 31, T11S, R5E (Stonefort field station 187). The faults ramp upward toward the south and are truncated at the base of the Battery Rock Sandstone (fig. 58). Therefore, the faults formed after deposition of the Wayside and before deposition of the Battery Rock. Potter (1957) described this exposure and attributed thrusting to gravitational slides off the flank of a rising McCormick Anticline. A possible angular unconformity between Wayside and Battery Rock Members was observed near the crest of the McCormick Anticline along a tributary of Caney Branch, SW SW NE, Section 11, T11S, R5E (Eddyville field station 402). Paleoslumps that create angular unconformity may be triggered by nontectonic causes, yet no such features have been found in the Caseyville in our study area away from the anticline.

Possible influence of the McCormick Anticline on distribution of sandstone in the Abbott Formation was mentioned in the discussion of environments of deposition of the Abbott. Inasmuch as the anticline is

nearly parallel to the dominant paleocurrent direction of lower Pennsylvanian strata, the case for contemporaneous tectonism is not particularly strong. The basal unit of the Abbott exposed at the north portal of the Illinois Central Railroad tunnel (see p. 25-26) was described as a "breccia" and attributed to subaqueous gravity slides off the flank of the anticline (Potter 1957). This deposit is not clearly a gravity slide; if it is a slide, movement might have been triggered by nontectonic causes. No similar rocks have been described elsewhere in the Illinois Basin.

No evidence was found for contemporaneous structural movements during deposition of the Spoon and Carbondale Formations or for any deformation on the New Burnside Anticline during Pennsylvanian sedimentation.

In summary, although initial uplift of the McCormick Anticline in Morrowan and Atokan time may have influenced sedimentation patterns and triggered gravitational slides, such movements, if they occurred, were much smaller than post-Atokan movements. Uplift of only a few feet, however, could divert the course of a deltaic distributary as accompanying earth tremors triggered slumping of water-saturated sediments.

Little Cache Fault Zone

Two faults along Little Cache Creek in the southwestern corner of the Creal Springs Quadrangle are part of a zone of north-trending faults first mapped by Weller and Krey (1939). The name Little Cache Fault Zone, introduced in this report, is from the creek whose course is largely controlled by the fault zone.

The two faults form a graben about 1,800 feet wide and trend slightly east of north (fig. 37). The western fault has 200 to 250 feet of throw where it exits in the study area. Eroded fault-line scarps on Caseyville sandstone are present in several places and indicate that the fault plane dips steeply east or is vertical. The fault plane was observed in a small draw west of Little Cache Creek in NE SW SW, Section 27, T11S, R3E (Nelson's field station 91). Here, the fault strikes N5°E and dips 80° to 85°E. Clay gouge and breccia are present; drag is confined to a narrow zone. The eastern fault is poorly exposed. Its throw increases southward to roughly 150 feet at the southern edge of the quadrangle. Available evidence indicates that the eastern fault, like the western, is a steep normal fault.

I have traced the Little Cache Fault Zone several miles south of the pre-

75

sent study area and have observed only high-angle normal faults.

Another fault that may be part of the Little Cache Fault Zone was mapped along Larkin Creek in Section 15, T11S, R3E, Creal Springs Quadrangle. The fault strikes about N10°E and has the east side downthrown about 45 feet. In an exposure south of Larkin Creek in NE SW SE, Section 15, the fault plane dips about 80°E and bears vertical striations, indicating dip-slip faulting. Shale dipping 60° to 80°E is juxtaposed with sandstone. About 30 feet north is another fault with a strike of N30°W and vertical dip and nearly horizontal slickensides. Sandstone adjacent to this fault dips northwest and is upthrown 5 to 6 feet on the southwest. The offset was right lateral (assuming horizontal movement). The strike-slip fault has an orientation compatible with a conjugate dextral fault produced in a stress field with maximum stress N10°E, minimum stress N80°W, and intermediate stress vertical. Normal faults that strike N10°E would form under the same stress conditions.

The Little Cache Fault Zone probably is a product of the episode of extensional tectonics that caused normal faulting elsewhere in the study area.

Other Structures

Faults in northeastern Creal Springs Quadrangle

Two small faults apparently unrelated to other structures in the area have been mapped in the northeastern part of the Creal Springs Quadrangle: one in N1/2, Section 4, T11S, R4E, and the other in Section 28, T10S, R4E. These faults are mapped on the basis of vertical offset of coals and other marker beds in the Spoon Formation. Both strike north-northwest and have the northeast side downthrown approximately 25 feet. The type of fault could not be determined but probably is normal.

Igneous dikes

Ultramafic dikes were uncovered late in 1984 during surface mining for coal at Peabody Coal Company's Will Scarlet Mine, just north of the Eddyville Quadrangle in Sections 13 and 14, T10S, R5E, Saline County. The largest dike was 10 to 23 feet wide and had a strike of N30°W and vertical dip. Several parallel dikes a few inches to 3 feet wide were nearby. The igneous

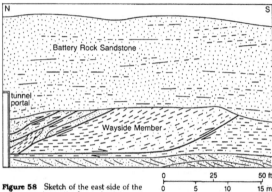

Figure 58 Sketch of the east side of the railroad cut immediately south of the tunnel portal in SE NW NE, Section 31, T11S, R5E, Stonefort Quadrangle, showing a series of imbricate north-dipping thrust faults within the Wayside Member of the Caseyville Formation. The faults steepen upward and are truncated at the base of the overlying Battery Rock Sandstone Member, indicating that faulting took place shortly after deposition of Wayside strata and before deposition of Battery Rock.

rock is dark green to nearly black, finely crystalline to porphyritic, and contains numerous inclusions of sedimentary rocks. It was weathered to ocher-colored clay as deeply as 30 feet below top of sedimentary rock. On the basis of petrographic study, the rock is an alnoite, composed of altered olivine phenocrysts in a ground mass of melilite, spinel, mica, perovskite, and carbonate. The rock has a globular segregationary (or fragmental) texture (D. G. Fullerton, Exmin Corporation, written communication, 1988). Similar compositions are reported for other intrusive rocks in the region (Clegg and Bradbury 1956, Hook 1974, Trace 1974). One dike observed at the strip mine and not studied petrographically appeared to be dominantly carbonate. Contact zones several inches wide on the small dikes and up to 5 feet wide on the large dike consist of coked coal and baked, fractured, and mineralized shale and siltstone.

Ground magnetometer surveys by Richard Lewis (personal communication, 1985) suggest that the large dike extends southeastward into the Eddyville Quadrangle. Lewis also predicted a sill or laccolith in Valmeyeran carbonate rocks in Section 25, T10S, R5E. Heigold and Robare's (1978) map shows a magnetic high here. Two oil-test holes in Section 25 bottomed in Chesterian strata without encountering igneous rock.

L. G. Henbest (ISGS, unpublished field notes, 1926) reported an igneous dike containing fragments of "granite, dolerite, and gneiss" in a ravine in E1/2 SE SW, Section 4, T11S, R6E. We found no igneous rock at this location (Eddyville field station 121); however, a rough-textured coarse breccia of sedimentary rock fragments in a matrix of bright orange-red material was found along a fault that strikes northwest and dips vertically. The outcrop could be a diatreme or explosion breccia similar to those around Hicks Dome (Baxter and Desborough 1965, Baxter et al. 1967). If Henbest's descriptions of inclusions are accurate, fragments of Precambrian rock have been blasted upward no less than 15,000 feet.

MINERAL DEPOSITS

Coal

Small-scale commercial and local coal mining has taken place within the study area from the early 20th century (and probably before) through the mid-1970s. Several coal seams in the Abbott and Spoon Formations locally reach a thickness of 3 to 4 feet and occur close enough to the surface for strip mining or shallow underground mining. These coals have among the highest heating values of any in the Illinois Basin (Damberger 1971) and some of them are low in sulfur. The main deterrent to mining is the sporadic distribution of the coal. The seams are largely lenticular and grade laterally into clastic rocks. Faults and folds further segment the deposits into small tracts. With extensive thick coals of the Carbondale Formation available just to the north, the study area is not attractive for large-scale mining under current economic conditions. However, opportunities exist for small surface mines and possibly drift or slope mines in several parts of the report area.

Coals of Caseyville Formation

The Gentry Coal Bed, near the base of the Drury Member, crops out in the southwestern part of the Stonefort Quadrangle. The Gentry and a slightly younger coal in the Drury Member are known from drilling and outcrops along the McCormick Anticline between Burden Falls and Delwood, but no coal thicker than 24 inches has been found. The Gentry Coal commonly contains a few inches of clean coal near the top, but the rest is thinly interlaminated shale and coal. A few caved-in adits or prospect pits in this coal have been seen; no commercial mining has been attempted or appears feasible.

Reynoldsburg Coal Bed

The Reynoldsburg Coal was mined for local use in the southeastern Creal Springs and southwesternmost Stonefort Quadrangles. West of Reynoldburg, in Sections 31 and 32, T11S, R4E, the coal varies from 15 to 36 inches thick, but rapid changes in thickness there render reserve estimates and mine planning difficult. However, the coal is, at least locally, of superior quality. Among the first eight samples listed in table 5, the average ash content was 4.3 percent, average sulfur 1.27 percent, and average heating value 13,750 Btu/lb. This compares with typical values of 8 to 12 percent ash, 2 to 5 percent sulfur, and 11,800 to 12,700 Btu/lb for coals of the Carbondale Formation in southern Illinois. The character and quality of the Reynoldsburg Coal vary laterally, as does the thickness. For example, the Reynoldsburg becomes canneloid and was tested for use as an oil shale along Ozark Creek, in Section 27, T11S, R4E (Barrett 1922). Also, the split Reynoldsburg Coal in borehole S-4 has high ash and extremely high sulfur content (table 5). At best, the Reynoldsburg presents opportunities for small-scale strip mining.

Tunnel Hill Coal Bed

The Tunnel Hill Coal is known only from a strip mine in NW NW, Section 6, T12S, R4E, and from test holes C-3 and C-5 (table 1), all in the southern part of the Creal Springs Quadrangle. Collapsed drift mines and test pits suggest wider distribution of the coal, but the coal itself was not seen. In the boreholes and strip mine, the Tunnel Hill varied from 12 to 28 inches thick. Two weathered samples of coal from the mine were analyzed and showed about 5 percent ash, 0.6 percent sulfur, and 10,400 Btu/lb; the heating value and probably the sulfur content were lowered by oxidation. Coal from the two drill cores was not analyzed because it was thin and contained many shale partings.

The Tunnel Hill Coal must be regarded as a marginal mining prospect. It might present a worthwhile target if found together with minable Reynoldsburg Coal. However, the Tunnel Hill appears to be thin or absent where the Reynoldsburg is well developed.

Oldtown Coal Bed

The Oldtown Coal Bed has been mapped in the Stonefort Quadrangle, mainly along the northern edge of T11S, and extends a short distance into the adjacent Eddyville and Creal Springs Quadrangles. The coal was surface mined just west of the railroad tracks near Oldtown and was taken from small drifts or "dogholes" elsewhere. It is 30 to 36 inches thick in the surface mine and railroad cut; Smith (1957) reported 42 inches at Frank Durfee's strip mine at the western edge of the Eddyville Quadrangle. Only a few inches of coal were found in test holes S-2 and S-3 (table 1) and in test holes drilled a short distance west of Durfee's mine. The "golden sandstone" generally directly overlies or occurs within a few feet of the top of the Oldtown Coal; the coal probably is eroded in many places.

Analytical results of the Oldtown Coal from the E. and L. Coal Company strip mine (table 5) indicate moderate ash and low sulfur content. Analysis of the coal from test hole S-3, where the seam was only 0.4 feet thick, probably is not representative of the coal where it reaches minable thickness. Like coals mentioned above, the Oldtown Coal may be locally recoverable by surface mining.

Delwood Coal Bed

The Delwood Coal Bed probably offers the best prospects for mining of any coal seam in the study area. It is present in all three quadrangles and is 3 to 4 feet thick over much of its extent. The largest known deposit lies in the triangular structural depression that includes the type locality in Sections 33 and 34, T10S, R6E, and Sections 3 and 4, T11S, R6E. It also occurs near the northern edge of the quadrangle in parts of Sections 27 and 28, T10S, R6E. These two areas together comprise about 525 acres. Within them the coal ranges from about 24 to 47 inches thick. Assuming an average thickness of 36 inches, we estimate a total resource of approximately 2.8 million tons of coal, all within 75 feet of the ground surface. A drawback to mining this coal is the common presence of a clay parting several inches thick near the middle of the seam. No analyses of the Delwood Coal in the Eddyville Quadrangle are available.

The Delwood Coal is known in the Stonefort Quadrangle only from the core of hole S-2. In this hole, the coal was 42 inches thick, with a 2-inch claystone parting near the middle; its depth was 23 feet. Most of the shallow reserves probably underlie the marshy bottomlands of Pond Creek. Northwest of Pond Creek, the coal is far below drainage. The Delwood Coal lies approximately 110 feet below the Mt. Rorah Coal, the outcrop of which is plotted on the geologic map. Results of chemical analyses for the Delwood Coal from the Stonefort Quadrangle (hole S-2) are given in table 5.

The Delwood Coal is widespread in the northern part of the Creal Springs Quadrangle. Its outcrop has been traced for several miles in the area north and west of New Burnside village, and the coal probably occurs in the subsurface farther north and west. The thickness of the coal varies from 24 to 72 inches, and a claystone parting up to 18 inches thick is commonly present. The Delwood Coal probably lies at strippable depth over several square miles, but no attempt has been made to calculate its tonnage because of the paucity of control points. No chemical analyses are available for the Delwood Coal in the Creal Springs Quadrangle.

New Burnside Coal Bed

The New Burnside Coal underlies most of Section 7 and parts of Sections 4, 5, 6, and 8, T11S, R4E, Creal Springs Quadrangle. The New Burnside Coal is largely coextensive with the underlying Delwood Coal and has been surface mined with it. The thickness varies from a few inches to 54 inches. Exposures inb strip mines (fig. 20) indicate that rapid lateral thickness changes, splitting, and local erosion are common with this seam. Therefore, closely spaced test drilling is needed to characterize its reserves and extent. Chemical analysis (table 5) indicate moderate ash and sulfur content and a heating value of nearly 13,000 Btu/lb.

Murphysboro Coal Member Coal believed to be the Murphysboro Coal occurs in a small part of E1/2, Section 8, T11S, R4E, Creal Springs Quadrangle. The seam is not currently exposed, but several collapsed mine adits suggest that it is at least a couple of feet thick. The Murphysboro occurs 20 to 30 feet above the New Burnside Coal, and the two coals probably could be strip mined together. Elsewhere in the study area, outcrops and borehole data show the Murphysboro to be thin or absent.

Mt. Rorah and Wise Ridge Coals
The Mt. Rorah Coal Member underlies several square miles in the northeastern part of the Creal Springs Quadrangle and northwestern part of the Stonefort Quadrangles and has been exploited in numerous small surface and underground mines. The thickness in this area appears to be consistently about 3 feet. Its gentle dip and topographic position should allow widespread contour and area surface mining, but the quality of the coal is poor. Claystone and shale partings are common, numerous claystone dikes were observed in Mt. Rorah Coal near Brushy Creek, and the sulfur content is high (table 5). As with all coals of the study area, additional drilling and sampling will be required to assess the minability of the Mt. Rorah Coal. The Wise Ridge Coal Bed occurs a short distance above the Mt. Rorah in most of the study area, but has not been observed thicker than 12 inches.

Coals of Carbondale Formation
The Carbondale Formation in the study area includes the Davis, Dekoven, Colchester (No. 2), Survant, and Houchin Creek (formerly Summum

(No. 4)) Coal Members. A small area of Carbondale Formation is mapped northwest of the abandoned Brown Brothers Excavating Company No. 1 Mine in Section 30, T10S, R6E, Eddyville Quadrangle. The Brown Brothers mined the Davis, Dekoven, and Houchin Creek Coals and explored the remaining reserves northwest of the mine. The nature of the deposits is inadequately defined because of complex faulting and poor exposure.

Oil and Gas
A total of 29 oil and gas test wells are known to have been drilled: 9 each in the Eddyville and Stonefort Quadrangles and 11 in the Creal Springs Quadrangle (table 3). Slight shows of oil or gas were noted in several wells, but none are known to have yielded commercial quantities of hydrocarbons, and all were abandoned.

All except three of these wells reached total depth in Mississippian rocks (table 3). Most of those drilled before 1950 were drilled with cable tools and finished in strata between the Palestine Sandstone and the Ste. Genevieve Limestone. Wells after 1950 were drilled with rotary rigs; most reached the Valmeyeran Ste. Genevieve or St. Louis Limestones. The Jenkins No. 1 Mohler (well 21, table 3) bottomed in lower Devonian Clear Creek Chert at a total depth of 4,250 feet. Two wells drilled by the Texas Pacific Company in the northeastern part of the Eddyville Quadrangle penetrated sub-Devonian strata. The John Wells et al. No. 1 (well 2, table 3) bottomed in Silurian carbonate at 6,200 feet, and the Mary Streich No. 1 Well (well 7, table 3) finished in upper Cambrian Mt. Simon Sandstone at 14,942 feet. To date, this well is the deepest drilled in Illinois.

Most exploration focused on obvious structural targets: the McCormick and New Burnside Anticlines. One or more wells have been placed on all large areas of closure on both anticlines. Other wells were drilled on anticlinal flanks, presumably in a search for combination stratigraphic/ structural traps. A few wells are in synclines. Only two have been drilled south of the McCormick Anticline.

Minor production of oil and gas has been achieved north of the study area. The Mitchellsville field, north of the Eddyville Quadrangle in Sec-

tions 2 and 15, T10S, R6E, contains three oil wells and one shut-in gas well. Cumulative oil production through 1983 was about 27,000 barrels. One well each produces oil from the Degonia, Waltersburg, and Cypress Sandstones (Chesterian); the gas well was completed in the Waltersburg. Outside the Mitchellsville area, only small and widely scattered hydrocarbon deposits have been found south of the Cottage Grove Fault System in Saline and Williamson Counties. Most wells had initial productions of 20 to 80 BOPD and quickly declined to less than 10 BOPD.

That exploratory success has been so limited is difficult to explain. Some operators have noted the lack of porosity in reservoir rocks and suggested that pore space was lost because of mineralization or recrystallization. Petrographic studies are required to test this idea. As for why the big anticlines have not yielded oil, perhaps the many faults and fractures allowed it to escape from shallow reservoirs. Seismic exploration is needed to evaluate internal structure of these folds, which may not resemble surface structure.

Fluorspar and Related Minerals
Fluorspar was mined at the Lost 40 Mine in the Eddyville Quadrangle and at several properties along the Lusk Creek Fault Zone adjacent to the study area. Small amounts of galena, sphalerite, and barite are found along with fluorspar (Weller et al. 1952). The Lost 40 and other mines were small, shallow operations that exploited ore in Pennsylvanian and upper Chesterian wall rocks. Elsewhere in the Illinois-Kentucky fluorspar district, the richest vein deposits are generally found in lower Chesterian units and in the upper part of the Ste. Genevieve Limestone. Thus, prospects appear good for fluorspar along the Lusk Creek Fault Zone below the depths currently explored.

Lesser possibilities exist for fluorspar and related minerals along the McCormick and New Burnside Anticlines. Normal faults in the McCormick and New Burnside Anticlines apparently have common origin with mineralized faults in the fluorspar district to the southeast. The critical question is whether faults on the McCormick and New Burnside structures were accessible to the same mineralizing fluids. If these are reactivated detached thrust faults, as we believe, they would connect with ore-solution pathways only via a series of dêcollements in ductile shales. Movement of mineralizing fluids along such decollements seems unlikely.

Stone, Sand, and Gravel

Stone, sand, and gravel occur within the study area, but distance to market and lack of transportation facilities render commercial development unlikely. However, some of these materials are suitable for local use.

The Kinkaid Formation contains limestone that may be suitable for fill material, road gravel, riprap, or agricultural use. The basal Negli Creek Member, which averages about 30 feet thick, is mostly limestone, but is argillaceous and cherty.

The Cave Hill Member probably contains too much shale to be worth quarrying in the study area. The Goreville Limestone Member reaches 45 feet thick and is less shaly and cherty than the rest of the Kinkaid. It has been quarried in the Waltersburg Quadrangle less than 1 mile south of the study area. Nearly the entire thickness of the Kinkaid Limestone is quarried on a large scale near Buncombe in Johnson County, about 6 miles southwest of the Creal Springs Quadrangle. The Kinkaid is less shaly at Buncombe than in the study area. The Buncombe quarry also has the advantage of being located along a railroad. The small deposits of Kinkaid Limestone within the study area are remote from railroads or good roads.

Sandstone occurs throughout the three quadrangles, but similar sandstone is abundant throughout southern Illinois. In the study area, the only likely use for sandstone is as fill for construction projects on location.

Sand and gravel deposits are small, thin, and sporadic. Stream deposits are largely silt and clay, with intermixed sand and rock fragments. Local deposits of sand from weathered sandstone occur at such sites as Sand Hill in the northern part of the Stonefort Quadrangle. These sands might be usable in small amounts as fill on location. Much better sand and gravel deposits, convenient to transportation, are being exploited in the Mississippi Embayment south of the study area and in the region of glacial drift to the north.

79

REFERENCES

Abegg, F. E., 1986, Carbonate petrology, paleoecology, and depositional environments of the Clore Formation (Upper Chesterian) in southern Illinois: M.S. thesis, Southern Illinois University, Carbondale, 222 p.

Amos, D. H., 1965, Geologic map of the Shetlerville and Rosiclare Quadrangles, Livingston and Crittenden Counties, Kentucky: U.S. Geological Survey, Geologic Quadrangle Map GQ-400.

Amos, D. H., 1966, Geologic map of the Golconda and Brownfield Quadrangles, Livingston County, Kentucky: U.S. Geological Survey, Geologic Quadrangle Map, GQ-546.

Atherton, E., 1971, Tectonic development of the eastern interior region of the United States: Illinois State Geological Survey, Illinois Petroleum 96, p. 29-43.

Barrett, N. V., 1922, Notes on Illinois bituminous shales, including results of their experimental distillation: Illinois State Geological Survey, Bulletin 38F, 20 p.

Baxter, J. W., and G. A. Desborough, 1965, Areal geology of the Illinois fluorspar district. Part 2 - Karbers Ridge and Rosiclare Quadrangles: Illinois State Geological Survey, Circular 385, 40 p.

Baxter, J. W., G. A. Desborough, and C. W. Shaw, 1967, Areal geology of the Illinois fluorspar district. Part 3 - Herod and Shetlerville Quadrangles: Illinois State Geological Survey, Circular 413, 41 p.

Baxter, J. W., P. E. Potter, and F. L. Doyle, 1963, Areal geology of the Illinois fluorspar district: Part 1, Saline Mines, Cave-in-Rock, De Koven, and Repton Quadrangles: Illinois State Geological Survey, Circular 342, 43 p.

Beckwith, R. H., 1941, Trace-slip faults: American Association of Petroleum Geologists Bulletin, v. 25, no. 12, p. 2181-2193.

Bell, W. A., 1938, Fossil flora of Sydney coalfield, Nova Scotia: Geological Survey of Canada, Memoir 215, 334 p.

Bristol, H. M., and R. H. Howard, 1971, Paleogeologic map of the sub-Pennsylvanian Chesterian (upper Mississippian) surface in the Illinois Basin: Illinois State Geological Survey, Circular 458, 16 p.

Brokaw, A. D., 1916, Preliminary oil report on southern Illinois - parts of Saline, Williamson, Pope, and Johnson Counties: Illinois State Geological Survey, Bulletin 35, p. 19-37.

Brown, L. F., A. W. Cleaves, and A. W. Erxleban, 1973, Pennsylvanian depositional systems in north-central Texas: a guide for interpreting terrigenous clastic facies in a cratonic basin: University of Texas at Austin, Bureau of Economic Geology, Guidebook 14, 122 p.

Buchanan, D. M., 1985, Carbonate petrology of the Negli Creek Limestone Member, Kinkaid Formation (Chesterian) in southern Illinois: M.S. thesis, Southern Illinois University, Carbondale, 61 p.

Butts, C., 1925, Geology and mineral resources of the Equality-Shawneetown area (parts of Gallatin and Saline Counties): Illinois State Geological Survey, Bulletin 47, 76 p.

Cady, G. H., 1926, The areal geology of Saline County: Illinois Academy of Science Transactions, v. 19, p. 250-272.

Cecil, C. B., and others, 1985, Paleoclimate controls on late Paleozoic sedimentation and peat formation in the central Appalachian basin (U.S.A.): International Journal of Coal Geology, v. 5, no. 2, p. 195-230.

Chamberlain, C. K., 1971, Morphology and ethology of trace fossils from the Ouachita Mountains, southeastern Oklahoma: Journal of Paleontology, v. 45, no. 2, p. 212-246.

Clegg, K. E., and J. C. Bradbury, 1956, Igneous intrusive rocks in Illinois and their economic significance: Illinois State Geological Survey, Report of Investigations 197, 19 p.

Cox, E. T., 1875, Geology of Gallatin and Saline Counties, in A. H. Worthen, director, Geological Survey of Illinois, v. VI, p. 197-219.

Damberger, H. H., 1971, Coalification pattern of the Illinois Basin: Economic Geology, v. 66, p. 488-494.

Desborough, G. A., 1961, Geology of the Pomona Quadrangle, Illinois: Illinois State Geological Survey, Circular 320, 16 p.

Devera, J. A., in preparation, Geology of the Glendale Quadrangle, Johnson and Pope Counties, Illinois: Illinois State Geological Survey, map and report.

Devera, J. A., C. E. Mason, and R. A. Peppers, 1987, A marine shale in the Caseyville Formation (Lower Pennsylvanian) in southern Illinois: Geological Society of America, Abstracts with Programs, p. 220.

Douglass, R. C., 1987, Fusulinid biostratigraphy and correlations between the Appalachian and Eastern Interior Basins: U.S. Geological Survey, Professional Paper 1451, 95 p., 20 plates.

Drozdzewski, G., O. Bornemann, E. Kunz, and V. Wrede, 1980, Beitrage zur Tiefentektonik des Ruhrkarbons: Geologisches Landesamt Nordrhein-Westfalen, Krefeld, 192 p.

Englund, K. J., J. F. Windolph, Jr., and R. E. Thomas, 1986, Origin of thick, low-sulfur coal in the Lower Pennsylvanian Pocahontas Formation, Virginia and West Virginia, in P. C. Lyons and C. L. Rice, editors, Paleoenvironmental and tectonic controls in coal-forming basins of the United States: Geological Society of America, Special Paper 210, p. 49-62.

Ervin, C. P., and L. D. McGinnis, 1975, Reelfoot Rift: reactivated precursor to the Mississippi Embayment: Geological Society of America, Bulletin, v. 86, p. 1287-1295.

Ethridge, F. G., S. L. Leming, and D. A. Keck, 1975, Depositional environments of lower Pennsylvanian detrital sediments of southern Illinois, *in* F. G. Ethridge, G. Reaungelter, and J. Utgaard, editors, Depositional Environments of Selected Lower Pennsylvanian and Upper Mississippian sequences of Southern Illinois, 37th Annual Tri-State Field Conference Guidebook, Department of Geology, Southern Illinois University, Carbondale, 5-7 October 1973, p. 16-34.

Fehrenbacher, J. B., 1964, Soil survey of Johnson County, Illinois: University of Illinois, Agricultural Experiment Station, Soil Report 82, 72 p.

Fehrenbacher, J. B., and R. T. Odell, 1959, Soil survey of Williamson County, Illinois: University of Illinois, Urbana-Champaign, Agricultural Experiment Station, Soil Report 79, 72 p.

Follmer, L.R., J. P. Tandarich, and R. G. Darmody, 1985, The evolution of pedologic and geologic profile concepts in the Midcontinent, U.S.A.: Agronomy Abstracts 1985, American Society of Agronomy, Madison, Wisconsin, p. 191.

Frye, J. C., A. B. Leonard, H. B. Willman, and H. D. Glass, 1972, Geology and paleontology of the Late Pleistocene Lake Saline, southeastern Illinois: Illinois State Geological Survey, Circular 471, 44 p.

Frye, J. C., H. D. Glass, and H. B. Willman, 1962, Stratigraphy and mineralogy of the Wisconsinan loesses in Illinois: Illinois State Geological Survey Circular 334, 55 p.

Gori, P. L., and W. W. Hays, editors, 1984, Proceedings of the Symposium on the New Madrid Seismic Zone: U.S. Geological Survey, Open-File Report 84-770, 468 p.

Graham, R. C., 1985, The Quaternary history of the upper Cache River valley, southern Illinois: M.S. thesis, Southern Illinois University, Carbondale, 236 p.

Grogan, R. M. and J. C. Bradbury, 1968, Fluorite-zinc-lead deposits of the Illinois-Kentucky mining district, *in* J. D. Ridge, editors, Ore Deposits of the United States, 1933-1967, v. 1: AIME, New York, p. 370-399.

Hallberg, G. R., T. E. Fenton, and G. A. Miller, 1978a, Standard weathering zone terminology for the description of Quaternary

sediments in Iowa, *in* G. A. Hallberg, editor, Standard Procedures for the Evaluation of Quaternary Material of Iowa: Iowa Geological Survey, Technical Information Series, no. 8, p. 75-109.

Hallberg, G. R., J. R. Lucas, and C. M. Goodman, 1978, Semi-quantitative analysis of clay mineralogy, *in* G. A. Hallberg, editor, Standard procedures for the Evaluation of Quaternary Material of Iowa: Iowa Geological Survey, Technical Information Series, no. 8, p. 5-22.

Heigold, P. C., and P. L. Robare, 1978, A ground magnetic survey of the Johnson County area of Illinois: unpublished report to the U.S. Nuclear Regulatory Commission, 20 p.

Heinrich, P. V., 1982, Geomorphology and sedimentology of the Pleistocene Lake Saline, southern Illinois: M.S. thesis, University of Illinois, Urbana-Champaign, 144 p.

Henbest, L. G., 1928, Fusulinellas from the Stonefort Limestone Member of the Tradewater Formation: Journal of Paleontology, v. 2, no. 1, p. 70-85.

Henderson, E. D., in prep., Geologic map of the Quaternary Geology of the Eddyville 7.5-Minute Quadrangle, Illinois.

Henderson, E. D., 1987, Stack unit mapping and Quaternary history of the Eddyville 7.5-Minute Quadrangle, Southern Illinois: M.S. thesis, Southern Illinois University, Carbondale, 84 p.

Hildenbrand, T. G., M. F. Kane, and W. Stauder, 1977, Magnetic and gravity anomalies in the northern Mississippi Embayment and their spatial relation to seismicity: U.S. Geological Survey, Miscellaneous Field Studies Map MF-914.

Hook, J. W., 1974, Structure of the fault systems in the Illinois-Kentucky fluorspar district, *in* D. W. Hutcheson, editor, A Symposium on the Geology of Fluorspar: Kentucky Geological Survey Series X, Special Publication 22, p. 77-86.

Horberg, C. L., 1950, Bedrock topography of Illinois: Illinois State Geological Survey, Bulletin 73, 111 p.

Howard, R. H., 1979, The Mississippian-Pennsylvanian unconformity in the Illinois Basin - old and new thinking, *in* J. E. Palmer and R. R. Dutcher, editors, Depositional and Structural History of

the Illinois Basin, Part 2, Invited Papers: Ninth International Congress of Carboniferous Stratigraphy and Geology, Field Trip 9, Illinois State Geological Survey, Guidebook 15a, p. 34-42.

Howe, J. R., and T. L. Thompson, 1984, Tectonics, sedimentation and hydrocarbon potential of Reelfoot Rift: Oil and Gas Journal, 12 November 1984, p. 174-190.

Hughes, W. B., 1987, The Quaternary history of the Lower Cache valley, southern Illinois: M.S. thesis, Southern Illinois University, Carbondale, 181 p.

Illinois Department of Public Health, 1966, Rules and regulations for refuse disposal sites and facilities: Illinois Department of Public Health, Sanitary Engineering Division, Springfield, 7 p.

Jacobson, R. J., 1983, Murphysboro Coal, Jackson and Perry Counties: resources with low to medium sulfur potential: Illinois State Geological Survey, Illinois Mineral Notes 85, 19 p.

Jacobson, R. J., 1987, Stratigraphic correlation of the Seelyville, Dekoven, and Davis Coals of Illinois, Indiana, and Western Kentucky: Illinois State Geological Survey, Circular 539, 27 p.

Jacobson, R. J., in preparation, Geologic map of the Goreville Quadrangle, Johnson and Williamson Counties, Illinois: Illinois State Geological Survey, Champaign.

Jacobson, R. J., and C. B. Trask, 1983, New Burnside "Anticline" - part of Fluorspar Area Fault Complex? Abstract: 12th Annual Meeting, Eastern Section, American Association of Petroleum Geologists, Carbondale, Illinois.

Jacobson, R. J., C. B. Trask, C. H. Ault, D. D. Carr, H. H. Gray, W. A. Hasenmueller, D. Williams, and A. D. Williamson, 1985, Unifying nomenclature in the Pennsylvanian System of the Illinois Basin: Transactions of the Illinois State Academy of Science, v. 78, no. 1-2, p. 1-11; Illinois State Geological Survey, Reprint 1985K.

Jacobson, R. J., C. B. Trask, and R. D. Norby, 1983, A Morrowan-Atokan limestone from the lower Abbott Formation of southern Illinois: Geological Society of America, Abstracts with Programs, v. 15, no. 6, p. 602-603.

Jennings, J. R., and G. H. Fraunfelter, 1986, Preliminary report on

macropaleontology of strata above
and below the upper boundary of
the type Mississippian: Transactions of the Illinois Academy
of Science, v. 79, no. 3-4, p.
253-261.
Jerzykiewicz, T., and A. R. Sweet,
1986, Caliche and associated
impoverished palynological as-
semblages: an innovative line of
paleoclimatic research onto the
uppermost Cretaceous and Paleo-
cene of southern Alberta, in Current Research, Part B: Geological
Survey of Canada, Paper 86-1B, p.
653-663.
Johnson, R. W., C. Haygood, T. G.
Hildenbrand, W. J. Hinze, and P.
M. Kunselman, 1980, Aeromagnetic map of the east-central midcontinent of the United States: U.S.
Nuclear Regulatory Commission,
NUREG/CR-1662, 12 p.
Kehn, T., 1974, Geologic map of the
Dekoven and Saline Mines Quadrangles, Crittenden and Union
Counties, Kentucky: U.S. Geological Survey and Kentucky Geolog-
ical Survey Map GQ-1147.
Kempton, J. P., 1981, Three-dimen-
sional geologic mapping for
environmental studies in Illinois:
Illinois State Geological Survey,
Environmental Geology Notes
100, 43 p.
King, Philip B. and Helen M. Beik-
man, compilers, 1974, Geologic
map of the United States: U.S.
Geological Survey, (scale
1:2,500,000).
Klasner, J. S., 1982, Geologic map of
the Lusk Creek roadless area,
Pope County, Illinois: U.S. Geological Survey Miscellaneous Field
Studies Map MF-1405-A.
Klasner, J. S., 1983, Geologic map of
Burden Falls roadless area, Pope
County, Illinois: U.S. Geological
Survey Miscellaneous Field Stud-
ies Map MF-1565-A.
Koeninger, C. A., and C. F. Mans-
field, 1979, Earliest Pennsylvanian
depositional environments in cen-
tral southern Illinois, in J. E.
Palmer and R. R. Dutcher, editors,
Depositional and Structural His-
tory of the Illinois Basin, Part 2,
Invited Papers: Ninth Interna-
tional Congress of Carboniferous
Stratigraphy and Geology, Field
Trip 9, Illinois State Geological
Survey, Guidebook 15a, p. 76-80.
Kolata, D. R., and W. J. Nelson, in
press, Tectonic history of Illinois
Basin, in M. W. Leighton, D. R.

Kolata, D. F. Oltz, and J. J. Eidel,
editors, Interior Cratonic Basins:
American Association of Petro-
leum Geologists Memoir (World
Petroleum Basins), Tulsa, Okla-
homa.
Kolata, D. R., J. D. Treworgy, and
J. M. Masters, 1981, Structural
framework of the Mississippi Em-
bayment of southern Illinois:
Illinois State Geological Survey,
Circular 516, 38 p.
Kosanke, R. M., J. A. Simon, H. R.
Wanless, and H. B. Willman, 1960,
Classification of the Pennsylva-
nian strata of Illinois: Illinois State
Geological Survey, Report of
Investigations 214, 84 p.
Kulander, B. R., and S. L. Dean, 1986,
Structure and tectonics of central
and southern Appalachian Valley
and Ridge and Plateau Provinces,
West Virginia and Virginia: Ameri-
can Association of Petroleum
Geologists Bulletin, v. 70, no. 11,
p. 1674-1684.
Lamar, J. E., 1925, Geology and min-
eral resources of the Carbondale
Quadrangle: Illinois State Geological Survey, Bulletin 48, 172 p.
Lannon, M. S., 1989, The Quaternary
history and surficial geology of the
Stonefort 7.5-Minute Quadrangle,
southern Illinois: M.S. thesis,
Southern Illinois University, Carbondale, 192 p.
Lannon, M. S., in prep., Geologic
map of the Quaternary geology of
the Stonefort 7.5-Minute Quadrangle, Illinois.
Lee, W., 1916, Geology of the Shawneetown Quadrangle in Kentucky: Kentucky Geological Survey, Series 4, v. 4, part 2, 73 p.
Leighton, M. M., and J. A. Brophy,
1961, Illinoisan glaciation in Illinois: Journal of Geology, v. 69,
no. 1, p. 1-31.
Leighton, M. M., G. E. Ekblaw, and
L. Horborg, 1948, Physiographic
divisions of Illinois: Journal of
Geology, v. 56, no. 1, p. 16-53;
Illinois State Geological Survey,
Report of Investigations 129.
Lidiak, E. G., and I. Zietz, 1976,
Interpretation of aeromagnetic
anomalies between latitudes 37°N
and 38°N in the eastern and
central United States: Geological
Society of America, Special Paper
167, 37 p., 1 map.
Lineback, J. A., 1979, Quaternary
deposits of Illinois: Illinois State
Geological Survey, map (scale,
1:500,000).

Lines, E. F., 1912, Stratigraphy of
Illinois with reference to Portland-
cement resources: Illinois State
Geological Survey, Bulletin 17, p.
59-113.
Maples, C. G., and J. A. Waters, 1987,
Redefinition of the Meramecian/
Chesterian boundary (Mississip-
pian): Geology, v. 15, no. 7, July
1987, p. 647-651.
McGinnis, L. D., and P. C. Heigold,
1961, Regional maps of vertical
magnetic intensity in Illinois:
Illinois State Geological Survey,
Circular 324, 12 p.
McGinnis, L. D., P. C. Heigold,
C. P. Ervin, and M. Heidari, 1976,
The gravity field and tectonics of
Illinois: Illinois State Geological
Survey, Circular 494, 28 p.
McKay, E. D., 1979, Wisconsinan
loess stratigraphy of Illinois, in
Wisconsinan, Sangamonian, and
Illinoian stratigraphy in central
Illinois: Illinois State Geological
Survey, Guidebook 13, p. 95-108.
McKeown, F. A., and L. C. Pakiser,
editors, 1982, Investigations of the
New Madrid, Missouri, Earth-
quake Region: U.S. Geological
Survey Professional Paper 1236,
201 p.
Miles, C., and B. Weiss, 1978, Soil sur-
vey of Saline County, Illinois: Uni-
versity of Illinois at Urbana-Champaign, Agricultural Experiment
Station, Soil Report 102, 94 p.
Miller, M. F., 1984, Distribution of
biogenic structures in Paleozoic
nonmarine and marine-margin se-
quences: an actualistic model:
Journal of Paleontology, v. 58, no.
2, p. 550-570.
Nelson, W. J., in preparation, Geo-
logic map of the Bloomfield Quad-
rangle, Johnson County, Illinois:
Illinois State Geological Survey,
map and report.
Nelson, W. J., and H.-F. Krausse,
1981, The Cottage Grove Fault
System in southern Illinois: Il-
linois State Geological Survey,
Circular 522, 65 p.
Nelson, W. J., and D. K. Lumm, 1984,
Structural geology of southeastern
Illinois and vicinity: Illinois State
Geological Survey, Contract/Grant
Report 1984-2, 127 p.
Nelson, W. J., and D. K. Lumm, 1985,
The Ste. Genevieve Fault Zone,
Missouri and Illinois: Illinois State
Geological Survey, Contract/Grant
Report 1985-3, 94 p.
Nelson, W. J., and D. K. Lumm,
1986a, Geologic map of the Shaw-

neetown Quadrangle, Gallatin County, Illinois: Illinois State Geological Survey, Illinois Geologic Quadrangle Map Series 1 (1:24,000).

Nelson, W. J., and D. K. Lumm, 1986b, Geologic map of the Equality Quadrangle, Gallatin and Saline Counties, Illinois: Illinois State Geological Survey, Illinois Geologic Quadrangle Map Series 2 (1:24,000).

Nelson, W. J., and D. K. Lumm, 1986c, Geologic map of the Rudement Quadrangle, Saline County, Illinois: Illinois State Geological Survey, Illinois Geologic Quadrangle Map Series 3 (1:24,000).

Nelson, W. J. Lumm, D. K. 1987, Structural geology of southeastern Illinois and vicinity: Illinois State Geological Survey Circular 538, 70p., 2 plates.

Nelson, W. J. and D. K. Lumm, 1990a, Geologic map of the Eddyville Quadrangle, Illinois: Illinois State Geological Survey IGQ-5, (1:24,000).

Nelson, W. J. and D. K. Lumm, 1990b, Geologic map of the Stonefort Quadrangle, Illinois: Illinois State Geological Survey, IGQ-6, (1:24,000).

Noggle, K. S., 1986, Stratigraphy and structure of the Leavitt Reservoir Quadrangle, Bighorn County, Wyoming: M.S. thesis, Iowa State University, Ames, 97 p.

North American Stratigraphic Code, 1983, The North American Commission on Stratigraphic Nomenclature: American Association of Petroleum Geologists Bulletin, v. 67, no. 5, p. 41-875.

Nuttli, O. W., 1973, The Mississippi Valley earthquakes of 1811-12: intensities, ground motion and magnitudes: Bulletin of the Seismological Society of America, v. 63, no. 1, p. 227-248.

Oliver, L., 1988, The stack unit mapping, Quaternary stratigraphy, and engineering properties of the surficial geology of the Waltersburg 7.5-Minute Quadrangle, Pope County, Illinois: M.S. thesis, Southern Illinois University, Carbondale, 149 p.

Owen, D. D., 1856, Report on the geological survey in Kentucky made during the years 1854 and 1855: Kentucky Geological Survey Bulletin, Series I, v. 1, 416 p.

Parks, W., 1975, Soil Survey of Pope, Hardin, and Massac Counties,

Illinois: University of Illinois at Urbana-Champaign, Agricultural Experiment Station, Soil Report 94, 126 p.

Peppers, R. A., 1988, Palynological correlation of major Pennsylvanian (upper Carboniferous) time-stratigraphic boundaries in the Illinois Basin with those in other coal regions of Euramerica: Geological Society of America, Abstracts with Programs, v. 20, no. 5, p. 384.

Peppers, R. A., and J. T. Popp, 1979, Stratigraphy of the lower part of the Pennsylvanian System in southeastern Illinois and adjacent portions of Indiana and Kentucky, in J. E. Palmer and R. R. Dutcher, editors, Depositional and Structural History of the Illinois Basin, Part 2: Invited Papers: Ninth International Congress of Carboniferous Stratigraphy and Geology, Field Trip 9, Illinois State Geological Survey, Guidebook 15a, p. 65-72.

Petersen, F. A., 1983, Foreland detachment structures, in J. D. Lowell and R. Gries, editors, Rocky Mountain Foreland Basins and Uplifts: Rocky Mountain Association of Geologists, Denver, Colorado, p. 65-78.

Pfefferkorn, H. W., 1971, Note on Conostichus broadheadi Lesquereux (trace fossil: Pennsylvanian): Journal of Paleontology, v. 45, no. 5, p. 888-892.

Phillips, T. L., and R. A. Peppers, 1984, Changing patterns of Pennsylvanian coal-swamp vegetation and implications of climatic control on coal occurrence: International Journal of Coal Geology, v. 3, p. 205-255.

Potter, P. E., 1957, Breccia and small-scale Lower Pennsylvanian overthrusting in southern Illinois: American Association of Petroleum Geologists Bulletin, v. 41, no. 12, December 1957, p. 2695-2709, 12 figures.

Potter, P. E., 1963, Late Paleozoic sandstones of the Illinois Basin: Illinois State Geological Survey, Report of Investigations 217, 92 p.

Potter, P. E., and H. D. Glass, 1958, Petrology and sedimentation of the Pennsylvanian sediment in southern Illinois: a vertical profile: Illinois State Geological Survey, Report of Investigations 204, 60 p.

Potter, P. E., E. Nosow, N. M. Smith, D. H. Swann, and F. H. Walker,

1958, Chester cross-bedding and sandstone trends in Illinois Basin: American Association of Petroleum Geologists Bulletin, v. 42, no. 5, May 1958, p. 1013-1046.

Potter, P. E., and R. Siever, 1956, Sources of basal Pennsylvanian sediment in the Eastern Interior Basin, Part I, Cross-bedding: Journal of Geology, v. 64, no. 4, p. 317-335.

Randall, J. W., 1970, Environment of deposition and paleoecology of the Cave Hill Member, Kinkaid Formation (upper Mississippian) in southern Illinois: M.S. thesis, Southern Illinois University, Carbondale, 150 p.

Raoult, J. F., and F. Meilliez, 1987, The Variscan Front and the Midi Fault between the Channel and the Meuse River: Journal of Structural Geology, v. 9, no. 4, p. 473-480.

Rexroad, C. B., and G. K. Merrill, 1985, Conodont biostratigraphy and paleoecology of middle Carboniferous rocks in southern Illinois: Courier Forschung Institut Senckenberg, v. 74, p. 35-64.

Riggs, M. H., 1990, The surficial geologic mappping and Quaternary history of the Creal Springs Quadrangle, Southern Illinois: M.S. thesis, Southern Illinois University, Carbondale, 172 p.

Riggs, M. H., in prep., Geologic map of the Quaternary geology of the Creal Springs 7.5-Minute Quadrangle, Illinois.

Rodgers, J., 1963, Mechanics of Appalachian foreland folding in Pennsylvania and West Virginia: American Association of Petroleum Geologists Bulletin, v. 47, no. 8, p. 1527-1536.

Rust, B. R., M. R. Gibling, M. A. Best, S. J. Dilles, and A. G. Masson, 1987, A sedimentological overview of the coal-forming Morien Group (Pennsylvanian), Sydney Basin, Nova Scotia, Canada: Canadian Journal of Earth Sciences, v. 24, p. 1869-1885.

Schwalb, H. R., 1982, Paleozoic geology of the New Madrid area: U.S. Nuclear Regulatory Commission, NUREG CR-2909, 61 p.

Shaver, R. H., and others, 1986, Compendium of Paleozoic rock-unit stratigraphy in Indiana - a revision: Indiana Geological Survey Bulletin 59, 203 p.

Shaw, E. W., 1915, Newly discovered beds of extinct lakes in southern and western Illinois and adjacent states, in Year-Book for 1910: Administrative Report and Various Economic and Geological Papers: Illinois State Geological Survey, Bulletin 20, p. 139-157.

Shaw, E. W., and T. E. Savage, 1912, Description of the Murphysboro-Herrin Quadrangles: U.S. Geological Survey, Geology Atlas, Folio 185, 15 p.

Siever, R., 1951, The Mississippian-Pennsylvanian unconformity in southern Illinois: American Association of Petroleum Geologists Bulletin, v. 35, no. 3, p. 542-581.

Siever, R., and P. E. Potter, 1956, Sources of basal Pennsylvanian sediments in the Eastern Interior Basin; 2. Sedimentary petrology: Journal of Geology, v. 64, no. 4, p. 317-335.

Skehan, J. W., N. Rast, and S. Mosher, 1986, Paleoenvironmental and tectonic controls of sedimentation in coal-forming basins of southeastern New England, in P. C. Lyons and C. L. Rice, editors, Paleoenvironmental and tectonic controls in coal-forming basins of the United States: Geological Society of America, Special Paper 210, p. 9-30.

Smith, A. E., and J. E. Palmer, 1981, Geology and petroleum occurrences in the Rough Creek Fault Zone: some new ideas, in M. K. Luther, editor, Proceedings of the Technical Sessions, Kentucky Oil and Gas Association, 38th Annual Meeting, 6-7 June 1974: Kentucky Geological Survey, Series XI, Special Publication 3, p. 45-59.

Smith, W. H., 1957, Strippable coal reserves of Illinois: Part 1. Gallatin, Hardin, Johnson, Pope, Saline, and Williamson Counties: Illinois State Geological Survey, Circular 228, 39 p.

Soderberg, R. K., and G. R. Keller, 1981, Geophysical evidence for deep basin in western Kentucky: American Association of Petroleum Geologists Bulletin, v. 65, no. 2, p. 226-234.

Sonnefield, R. D., 1981, Geology of northwestern Jackson County, with special emphasis on the Caseyville Formation: M.S. thesis, Southern Illinois University, Carbondale, 84 p.

Stonehouse, H. B., and G. M. Wilson, 1955, Faults and other structures in southern Illinois—a compilation: Illinois State Geological Survey, Circular 195, 4 p.

Swann, D. H., 1963, Classification of Genevievian and Chesterian (Late Mississippian) rocks of Illinois: Illinois State Geological Survey, Report of Investigations 216, 91 p.

Thompson, M. W., R. H. Shaver, and E. A. Riggs, 1959, Early Pennsylvanian fusulinids and ostracods of the Illinois Basin: Journal of Paleontology, v. 33, no. 5, p. 770-792.

Trace, R. D., 1974, Illinois-Kentucky fluorspar district, in D. W. Hutcheson, editor, Symposium on the Geology of Fluorspar: Kentucky Geological Survey, Series X, Special Publication 22, p. 58-76.

Trace, R. D., and P. McGrain, 1985, The Chaetetella Zone in the Kinkaid Limestone (Mississippian); a useful stratigraphic marker along the southern rim of the Eastern Interior (Illinois) Basin: Kentucky Geological Survey, Series XI, Information Circular 14, 9 p.

Trask, C. B. and R. J. Jacobsen, 1990, Geologic map of the Creal Springs Quadrangle, Illinois: Illinois Geological Survey Map IGQ-4, (1:24,000.)

Walter, N. F., G. R. Hallberg, and T. E. Fenton, 1978, Particle size analysis by Iowa State University Soil Survey Laboratory, in G. A. Hallberg, editor, Standard Procedures for the Evaluation of Quaternary Material of Iowa: Iowa Geological Survey Technical Information Series, no. 8, p. 61-74.

Wanless, H. R., 1939, Pennsylvanian correlations in the Eastern Interior and Appalachian coal fields: Geological Society of America, Special Paper 17, 130 p.

Wanless, H. R., 1956, Classification of the Pennsylvanian rocks of Illinois as of 1956: Illinois State Geological Survey, Circular 217, 14 p.

Weibel, C. P., and W. J. Nelson, in preparation, Geologic map of the Lick Creek Quadrangle, southern Illinois: Illinois State Geological Survey, Champaign.

Weibel, P., J. Devera, and W. J. Nelson, in preparation, Geology of the Waltersburg Quadrangle, Pope County, Illinois: Illinois State Geological Survey, map and report.

Weller, S., 1913, Stratigraphy of the Chester Group in southwestern Illinois: Transactions of the Illinois State Academy of Science, v. 6, p. 118-129.

Weller, S., 1920, The Chester Series in Illinois: Journal of Geology, v. 28, no. 4, 5, p. 281-303; v. 28, no. 5, p. 395-416.

Weller, J. M., 1940, Geology and oil possibilities of extreme southern Illinois, Union, Johnson, Pope, Hardin, Alexander, Pulaski, and Massac Counties: Illinois State Geological Survey, Report of Investigations 71, 71 p.

Weller, S., and F. F. Krey, contributions by J. M. Weller, 1939, Preliminary geologic map of the Mississippian formations in the Dongola, Vienna, and Brownfield Quadrangles: Illinois State Geological Survey, Report of Investigations 60, 11 p.

Weller, J. M., R. M. Grogan, and F. E. Tippie, 1952, Geology of the fluorspar deposits of Illinois: Illinois State Geological Survey, Bulletin 76, 147 p.

Willman, H. B., and J. C. Frye, 1970, Pleistocene stratigraphy of Illinois: Illinois State Geological Survey, Bulletin 94, 204 p.

Willman, H. B., and J. C. Frye, 1980, The glacial boundary in southern Illinois: Illinois State Geological Survey, Circular 511, 23 p.

Willman, H. B., E. Atherton, T. C. Buschbach, C. Collinson, J. C. Frye, M. E. Hopkins, J. A. Lineback, and J. A. Simon, 1975, Handbook of Illinois stratigraphy: Illinois State Geological Survey, Bulletin 95, 261 p.

Wood, G. H., T. M. Kehn, and J. R. Eggleston, 1986, Depositional and structural history of the Pennsylvanian Anthracite region, in P. C. Lyons and C. L. Rice, editors, Paleoenvironmental and tectonic controls in coal-forming basins of the United States: Geological Society of America, Special Paper 210, p. 31-48.

Worthen, A. H., 1868, Geology of Alexander, Union, Jackson, Perry, Jersey, Greene, and Scott Counties, in A. H. Worthen, director, Geological Survey of Illinois, v. 3, p. 20-144.

Printed by authority of the State of Illinois/1991/1500

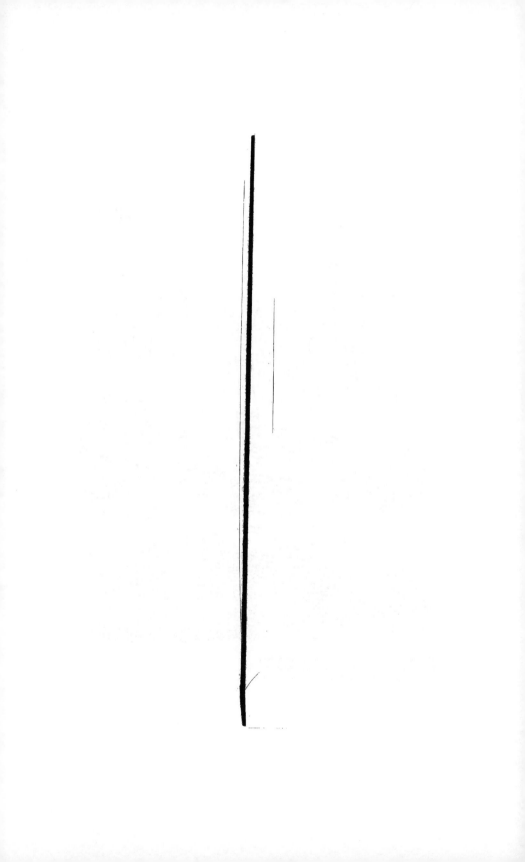

Murray Bluff Sandstone Member, upper Abbott
Formation, just east of the type locality. Strata dip
northwest (left) on flank of New Burnside Anticline.